T0137129

Modeling and Optimization in Science and Technologies

Volume 11

The book series *Modeling and Optimization in Science and Technologies (MOST)* publishes basic principles as well as novel theories and methods in the fast-evolving field of modeling and optimization. Topics of interest include, but are not limited to: methods for analysis, design and control of complex systems, networks and machines; methods for analysis, visualization and management of large data sets; use of supercomputers for modeling complex systems; digital signal processing; molecular modeling; and tools and software solutions for different scientific and technological purposes. Special emphasis is given to publications discussing novel theories and practical solutions that, by overcoming the limitations of traditional methods, may successfully address modern scientific challenges, thus promoting scientific and technological progress. The series publishes monographs, contributed volumes and conference proceedings, as well as advanced textbooks. The main targets of the series are graduate students, researchers and professionals working at the forefront of their fields.

More information about this series at http://www.springer.com/series/10577

Patricia Martin-Rodilla

Digging into Software Knowledge Generation in Cultural Heritage

Modeling Assistance Strategies for Large
Archaeological Data Sets

 Springer

Patricia Martin-Rodilla
Institute of Heritage Sciences (Incipit)
Spanish National Research Council (CSIC)
Santiago de Compostela
Spain

ISSN 2196-7326   ISSN 2196-7334 (electronic)
Modeling and Optimization in Science and Technologies
ISBN 978-3-319-88726-5   ISBN 978-3-319-69188-6 (eBook)
https://doi.org/10.1007/978-3-319-69188-6

Printed on acid-free paper

This Springer imprint is published by Springer Nature
The registered company is Springer International Publishing AG
The registered company address is: Gewerbestrasse 11, 6330 Cham, Switzerland

There are two lasting bequests we can give our children: Roots and Wings (Hodding Carter)

To my husband Miki, my parents and my sister Becky, for giving me both and hopefully transmitting them to the next generation very soon.

Foreword

Knowledge generation is either a trivial or a wicked problem. We all generate knowledge daily and automatically, so it should be affordable and quite inconsequential. Understanding how we do it, however, or why we generate the knowledge that we generate, constitutes a formidable challenge. This is especially so in the practice of the humanities, which are often based on interpretive processes that, on a superficial judgment, appear to work by magic.

In this book, Patricia focuses on scientific knowledge generated within the archaeological practice, describing what major mechanisms exist, what strategies we can employ to assist archaeologists in knowledge generation, and how well these strategies perform. In this regard, the book employs a wide range of research and narrative approaches, including exploratory exercises to uncover how archaeologists reason about evidences and ideas; discourse analysis to determine how inferences are anchored in narratives; conceptual modeling to organize discovered and proposed knowledge; method engineering to suggest a formal language to assist in these tasks; model-driven engineering to propose an architectural framework for assisted knowledge generation systems; software prototyping to construct a test bed for the posited hypothesis; and statistical approaches to analyze how well these approaches work in the real world.

In addition, this book constitutes one of the first large bodies of work that realizes the idea of co-research between software engineering and archaeology. In an age where Digital Humanities often offer more chaff than wheat, scholarly advances are needed that genuinely achieve bidirectional benefits between information technologies and the humanities. Through the work here presented, the author offers assistance to archaeologists in their knowledge generation processes and, at the same time, advances software engineering itself with specific proposals about systems architecture, conceptual modeling, and situational method engineering. This degree of transdisciplinarity is rarely attained.

I hope that this book helps the reader to construct a better understanding of the magic that takes place behind the scenes during archaeological interpretation. And, mostly, I hope it establishes the bases for what should be, in a not too distant future, a comprehensive treatment of knowledge generation in the humanities.

Cesar Gonzalez-Perez
Staff Scientist
Institute of Heritage Sciences (Incipit)
Spanish National Research Council (CSIC)
Santiago de Compostela, Spain

Preface

Archaeology constitutes a vast discipline involving several subdisciplines, theoretical positions, and research methods. Within this great variety, their professionals produce and manage large amounts of data about evidences of our past and present, from which they create new knowledge, constituting the knowledge about heritage of a particular community. This knowledge defines the community at present, and is transmitted to the present and future generations. In addition, any new piece of information or archaeological knowledge created serves as a basis to build new hypothesis of work or new research lines for an archaeological site, area, community, or heritage problem achieved. Thus, any archaeological data set captured, created, interpreted, and managed acquires special relevance, as basis of the archaeological knowledge that we produced, but, increasingly often, these data sets are large and complex information structures underestimated or used only in specific contexts like an archaeological site or project.

Current trends incorporate software engineering strategies from the beginning of the archaeological research conception in order to assist knowledge generation processes from large archaeological data sets. As a science of the information, software engineering provides a repository of techniques, methods, and tools to manage, process, use, and exploit information. Using them, it is possible to assist domain professionals in performing processes ranging from the analysis of raw gathered data to the generation of new knowledge based on them. Successful examples are software genomics research or business decision-making processes. This assistance is also possible in archaeology and heritage information.

However, and despite their relevance and the regular application of software engineering solutions to the archaeological domain, the knowledge generation process in archaeology poses a challenge for software engineering, mainly due to the lack of formal studies about it. This implies that we do not know which particular processes in archaeology we must assist and what should be the appropriate assistance in each case. Furthermore, the archaeological information, in general, the humanities, possesses some particular characteristics that are especially difficult to deal with by software, such as the presence of high subjectivity, the fact that much information is uncertain or vague, and the importance of the temporal aspect of the information.

All these aspects, among others, represent current challenges in archaeological software assistance. This book aims at presenting state-of-the-art research and future trends on software strategies to manage and build software systems focused on generating new knowledge from large archaeological data sets. As a result of years of research, the book also presents specific new techniques to integrate data conceptual and visual assistance in our archaeological software systems: how to model from archaeological textual sources, how to create dynamic and conceptually rich data models, how to capture cognitive processes performed in analyzing archaeological data sets, and how to define visual and interaction strategies to improve visualization areas in archaeological data analysis. All these parts are explained in detail and integrate in a conceptual framework that allows the reader to follow step by step the application to their own archaeological information.

The proposed conceptual framework has been validated through an archaeological case study, also implemented as a functional iOS system prototype. The prototype has been validated empirically by archaeologists, comparing the performance of knowledge generation processes using the proposed framework to the conventional ways without software assistance. The empirical validation reveals how the framework provides a robust solution for the construction of software systems to assist in the knowledge generation process in archaeology.

In light of these results, this book is an excellent source for students (as a reference for existing works and innovative literature in the topics addressed), for archaeologists and data managers in working on digital humanities domains (as a reference guide to apply software strategies in their daily analysis of archaeological data) and finally for researchers interested in how the archaeological knowledge is represented and generated, as well as how it is possible to assist these processes by software.

I hereby acknowledge to the people who helped me during research and this book completion. First, I would like to thank Dr. César González-Pérez at Incipit-CSIC for his unconditionally support and mentoring and to Prof. Óscar Pastor at PROS center, Universidad Politécnica de València, for his commitment to this work. I also thank the Spanish National Research Council for funding part of the research involved, and to all the related institutions in which I worked or I visited: University of Technology Sydney (Australia), Université Paris 1 Panthéon-Sorbonne (France), Andrés Bello University (Chile), Universidade Nova de Lisboa (Portugal), Incipit-CSIC, and PROS center. Finally, my thanks to Dr. Vicente Pelechano (Universitat Politècnica de València, Spain), Dr. Antonio Vallecillo (Universidad de Málaga, Spain), and Dr. Jeremy Huggett (University of Glasgow, Scotland) for their contributions and external evaluation of this work.

Santiago de Compostela, Spain   Patricia Martin-Rodilla
August 2017

Contents

List of Figures

Part I
Understanding Software Engineering and Archaeology Co-research

Chapter 1
Introduction

Software Engineering and Archaeology, So Far Away?

The disciplinary division which has traditionally been imposed in the form of humanities and social sciences versus natural sciences and engineering has not allowed for fluid two-way communication between the two worlds. In general, this connection has been produced by way of a hierarchical relationship of the disciplines, in such a way that any interdisciplinary study has typically had one discipline which takes on the main role and other, subsidiary, disciplines whose role is relegated to the application of the methods, techniques and tools which are of use to the main discipline. Nowadays, this situation is changing, thanks to the appearance of interdisciplinary studies which enable co-existence and collaboration at the same level between humanistic and engineering disciplines, such as computational linguistics and information documentation and retrieval, as well as natural and technological sciences, such as biotechnology and environmental studies.

Although the relation between software engineering and archaeology began decades ago (the major academic conference "Computer Applications and Quantitative Methods in Archaeology" goes back 45 years ago), the need to assist professionals working in archaeology in deep data analysis by software has already recently been identified, as can be seen in projects such as Europeana [78], Bamboo [24] and, more recently, Ariadne [17]. All these projects aiming to build infrastructures and software solutions for specific aspects in archaeology, identifying that software engineering plays a relevant role as an information science, since it brings together the corpus of knowledge necessary for the conceptualization, management, appropriate handling, exploitation and giving of assistance to researchers as they carry out their tasks and take decisions, from raw data to the generation of new knowledge. This is possible thanks to new approaches, techniques and sub-disciplines originating from the field, which allow for assistance to be provided by means of software in the afore-mentioned generation of knowledge. Thus, working with archaeological paradigms, processes and data allows software

© Springer International Publishing AG 2018
P. Martin-Rodilla, *Digging into Software Knowledge Generation in Cultural Heritage*, Modeling and Optimization in Science and Technologies 11, https://doi.org/10.1007/978-3-319-69188-6_1

engineering as a discipline to test its achievements, to detect its limitations and to undertake new lines of research which may arise from the characteristics of the new field.

In this context, attempting to assist humans in their knowledge generation process by software in a humanistic and social-based discipline as complex as archaeology could suggest a traditional approach as the basis for this study, in which archaeology as a discipline is assisted by methods, techniques and tools originating from software engineering. However, this reading of the topic that this book addressed is simplistic and is far from the true nature of the research which is put forward here. In fact, by dealing with this problem, we are affirming that the process of knowledge generation as an independent process from the field in which we are moving constitutes the true framework of the research which concerns us. In this context, the field in which this process of knowledge generation is set (archaeology) should be studied in depth, along with the field of software engineering as a co-existing field, due to the fact that both fields need to work in collaboration for the proposed assistance to be a success. It is for this reason that this book adopts a true interdisciplinary not only in approach, but also in terms of results, offering interesting strategies for different audiences and professionals from both worlds.

Next section goes on to illustrate through a real archaeological project what kind of processes and situations we can find when we are generating new knowledge about our past, based on which data and the specific characteristics of them.

Illustrating Knowledge Generation Based on Archaeological Data

Any archaeological project generally involves different disciplines and professionals with differing profiles and degrees of expertise as well as several research and management aims and organizations with different responsibilities and roles. This situation can mean that the data which is gathered and the information generated has a high degree of heterogeneity and, in most cases, there is a tendency towards ad hoc solutions which are restricted to one particular project. For example, in 2010 the Institute of Heritage Sciences in Spain carried out a research project entitled "Procesos de Patrimonialización no Camiño de Santiago: tramo Santiago-Fisterra-Muxía" (project financed by the Plan gallego de I+D+i (Incite), with the reference code INCITE09606181PR). The main aim of the project was to carry out multidisciplinary research on a section which is not officially recognized by the Catholic Church as being part of the Way of St James. This stretch, however, from Santiago de Compostela to Finisterre, is becoming more and more popular among some pilgrims, who do not stop in Santiago but go on to Finisterre. The aims of the project were to characterize the most representative places along this section of the Way, which is also known apocryphally as the "Way of the Atheists" [240], to find out why pilgrims do it and to gather information about the traditions associated with it and what socio-economic impact the increased popularity of this section has had. Thus it was necessary to carry out collaborative work between historians and archaeologists (who focused on the material evidence from the different sites excavated and studied), anthropologists (who focused on the study of the relationship between modern-day settlements and the different sites by way of interviews and ethnographic work) and sociologists (whose interest was centered on the touristic phenomenon of this section of the Way and its socio-economic impact on the various groups which were studied: tourists, local population, businesses in the region, etc.). After several fieldwork campaigns had been carried out by professionals from the three disciplines, the project produced semi-structured data (basically the results of records and cataloguing of archaeological evidence and the results of surveys) and data in an unstructured format (reports on archaeological research, technical reports of excavations, transcripts of interviews, archaeological and anthropological fieldwork diaries, etc.).

All of this raw data and information which was generated served as a basis to generate knowledge regarding our past. It allowed the professionals to establish, for example, the profile of the typical pilgrim who walks this section (if he/she is young or old, a believer or not, his/her motivation, if he/she walks alone or accompanied, etc.). It also allowed for elements of material heritage (hermitages, wayside crosses, etc.) to be catalogued for the first time on this section. It was, therefore, possible not only to gain isolated information about each element, but also to compare information about their relation to the stretch from Santiago to Finisterre and their usage, state of preservation and other parameters with other sites on the Way in the search for similarities and differences. But what is the process which allows for this generation of new knowledge?

If we consult the primary sources of the process, that is to say the researchers themselves who participated in the project, we discover that these results were obtained thanks to the conceptualization of a common framework of data to be gathered regarding the heritage elements, which allowed them to compare, correlate and search for contrasts (in other words, to carry out cognitive processes) regarding the usage, the state of preservation, the localization of the elements with respect to the Way and the results of the interviews. We shall also see that software was only used to support the spatial reasoning (Geographical Information Systems), and that manual drawings, diagrams and ad hoc visualizations were made in order to be able to visualize the data sets and to reason more deeply about the rest of the dimensions of the data. This allowed the data to be understood within its temporal context and meant that groups of data could be compared according to their characteristics (for example, the archaeological evidence and objects found) and where they appeared. An example of this manual outlining are the Harris diagrams [114], commonly referred to as the "Harris Matrix", which allow unearthed objects to be identified with their relative position in the excavation.

But can all this complex process be helped integrally by way of software? In what way? To what degree? For what kind of data? What happens in cases of large and complex data sets? Are there limitations? These informal questions form the initial motivation of this book.

Terminology Adoption

This section condenses the formal definition adopted of the most relevant terms used throughout this book. It is not a glossary of terms, but a clarification of the meaning and definition used in this work for terms that could constitute ambiguities or different scientific and / or disciplinary interpretations.

Knowledge generation: by the term knowledge generation we understand the process by which a human being, from the moment he/she comes into contact with descriptive data from real entities until, by way of cognitive processes based on analytical reasoning and "upward" changes in the level of abstraction, obtains previously unknown results regarding the analyzed entities. This process takes place at the core of what has been described by many authors [7, 27, 46] as a model of layers or levels, in which the ascent from one layer or another is subject to these cognitive processes being carried out by the human being.

Cultural Heritage: this is the cultural inheritance belonging to the past of a community, with which it co-exists in the present and which it passes from present to future generations. It includes the uses, representations, expressions, knowledge and techniques (as well as the instruments, objects, artefacts and cultural places which are inherent), which communities, groups and, in some cases, individuals recognize as part of their Cultural Heritage [264]. It should be pointed out here that, within UNESCO's definition, there are divergences when it comes to establishing criteria for defining the boundaries of the sphere of influence of Cultural Heritage [106]. In this book, we take this definition from a flexible perspective.

Software Assistance: by software-assisted, we understand the ensemble of services offered by a software system to human users to help them carry out certain (typically manual or well-defined) processes in any area. The term has become popular in certain fields, such as in software assistance for textual analysis. Good examples of this relate to the detection of plagiarism [254, 255, 283], industrial processes [230] and certain software-assisted medical processes, such as the management of hospital discharges and the detection of relevant elements in medical analyses [108, 227]. Throughout this book, in a similar way to other current studies [61], software-assisted processes will be addressed as an ensemble of services offered by a software system to specialists in a particular field (in this case archaeologists) in order to assist them in carrying out cognitive processes which will allow them to generate knowledge in their field.

Knowledge extraction: this refers to the ensemble of techniques which allow for the creation of new knowledge based on structured data (relational databases, XML) and unstructured data (text, documents, images). The resulting knowledge must be in a format which is legible for the machine but which, at the same time, enables the human to carry out cognitive or inference processes [263]. The techniques included in knowledge extraction are applied with differing degrees of automatization, from techniques based on ontologies [275] or annotation to the most automatic techniques related with the sub-discipline of the recuperation of information [23]. Throughout this book, the subgroup of these techniques which are manually and semi-automatically applied and are mainly based on domain ontologies and discourse analysis has been taken into account. However, the capacity of software-assisted processes for frameworks presented in the most automatic cases, such as the application to the discipline of data-mining techniques, the automatic recuperation of data and the automatic processing of natural language, has been considered as an area worthy of future study.

Information Visualization: this is taken to be an ensemble of techniques which "use visual computing to amplify human cognition with abstract information." [40]. It is "an increasingly important subdiscipline within HCI, (which) focuses on graphical mechanisms designed to show the structure of information and improve the cost of access to large data repositories" [21]. Over the course of this book, we shall look at existing techniques and creation mechanisms for new techniques within the discipline for their application to the visualization of archaeological data, proposing the most appropriate forms of visualization as the natural way of assisting people through software.

Book Outline

With the main idea of what knowledge generation process is, how we can use software to assist it when we focus on large archaeological data set and an illustrative example, we have set the stage for successive chapters. Before going into the different strategies and the overall proposal presented in this volume, Chap. 2 explains in detail the research methodology adopted during the entire book, and Chap. 3 reviews the existing techniques and tools that could serve us as a reference to assist archaeological knowledge generation by software. Based on that, we will discuss them and in Chaps. 4 and 5 we present some original techniques in order to deal with the archaeological data particularities and capture them for software needs.

Then, a major part of this volume is dedicated to explaining and guiding the reader through a conceptual framework that integrates modeling strategies to design and built software to assist archaeologist in analyzing their data. Chapters 6–10 present the framework proposed, while Chaps. 11 and 12 detail both the processes of evaluation and validation carried out on the framework created, each one being of a different nature; one is an analytical-formal validation and the other an empirical validation with archaeologists.

Finally, Part V contains chapters in which the impact, sphere of application and the implications of the results obtained throughout the book are addressed, as well as the future possibilities for each one of the original contributions presented, both from the point of view of the research community and from a practical perspective of their application by real users working on archaeological problems.

Thus, this book can be used in several ways depending on the main goal and the professional profile of the reader, highlighting:

- As scholar reference guide or complementary volume for existing works and innovative literature about software approaches to the knowledge generation and management of archaeological data, especially as part of archaeological education courses.
- For archaeologists and related professionals in conservation, preservation, archaeological education or technical support, scholars who are increasingly interested in emerging software strategies for researching with big archaeological data sets in their research daily agenda.
- For critical humanistic scholars interested in tracing the ways that the knowledge is represented and generated in archaeology.
- As conceptual and technical framework application manual for software/interface engineers and big data managers in working on digital humanities domains, especially archaeology, who are interested in visual aspects of the archaeological data.
- As conceptual and technical framework application manual for software engineers, archaeological specialists in data aspects and big data managers in working on digital humanities domains, especially archaeology, who are interested in more conceptual and data-based aspects of the archaeological domain. We can follow here how analyzing archaeological discourse and transform them into data, and how modeling using object oriented techniques archaeological data.

Each practical part of the book, especially conceptual and interface design techniques in Chaps. 4 and 5 and framework application in Part III contains examples in situ, as well as the reader can follow their application step by step in a real archaeological case study in Part IV.

Chapter 2
Prior Research Design

Initial Questions and Goals

The main objective of the research carried out here is determining and showing how it is possible for software to assist to the knowledge generation process as practiced by specialists in archaeology. Thus, we shall attempt at the beginning to identify possible improvements that can be made in the knowledge generation process by way of software assistance and to propose software models that can provide systems with such a capacity of assistance. This assistance could materialize in the form of guidelines for the application of knowledge extraction techniques or in the proposal of the application of techniques for the visualization of archaeological information, which are adapted to the characteristics of the knowledge-generation process. Due to the broad scope treated, this book focusses in the second type of software-assistance, that is, providing visualization techniques appropriated to the special characteristics of archaeological data sets and archaeologists' needs.

As any other research process, a fundamental research question must be asked in accordance with our objective:

To What Extent Is It Possible to Improve Knowledge Generation Processes in Archaeology by Way of Software Assistance to the User with Information Visualization Techniques?

Having defined each of the relevant concepts in the principal research question (see Terminology Adoption section in previous chapter), we can understand its scope more exactly. Taking as its starting point the principal research question, this study has the objective of validating the initial hypothesis on the premise that the answer to the question is affirmative. That is to say, we take the hypothesis that it is possible to significantly improve knowledge-generation processes in archaeology by way of providing software assistance to the user with information visualization techniques.

The fundamental research question posed here are too abstract to stablish a practical approach. Thus, we decomposed it in more specific secondary questions that allow us to deal with the complexity of them.

© Springer International Publishing AG 2018
P. Martin-Rodilla, *Digging into Software Knowledge Generation in Cultural Heritage*, Modeling and Optimization in Science and Technologies 11, https://doi.org/10.1007/978-3-319-69188-6_2

Firstly, and in order to prove that the software assistance is possible, it is necessary to know in depth how knowledge is generated in Archaeology. We wonder: **What problems exist in knowledge-generation processes in archaeology as they are normally carried out?**

Secondly, the study of the knowledge-generation process in the discipline allowed us to detect the need to formally characterize the cognitive processes carried out in the aforementioned process in order to improve it. Then, we need to answer to: **What are the most common cognitive processes carried out by archaeologists in the generation of knowledge?**

Finally, the application of information visualization techniques was included within the main research question in order to produce the software assistance, which we wish to provide. In this point, the need arose to study which of these techniques would offer us the software assistance we proposed. We need to know: **Which are some appropriate information visualization techniques to assist each one of the cognitive processes identified within archaeology?**

Along the book, we verify this affirmation by means of hypothetico-deductive reasoning, gathering evidence of all types, both formal and empirical, in favor of and against it. Although next section may deviate slightly from the central theme of the book for more practical readers, we believe it is necessary to document the research methodology employed, for two main reasons: its inherent relationship with the strategies and solutions developed here and explained in successive chapters, as well as to consider this methodology a reference for researchers interested in knowledge generation areas and what kind of underlying methodology is possible to follow for future developments.

"Design Science Methodology" in Practice

In order to answering the research questions raised, confirming and testing the possible improvements via software to the knowledge generation process in archaeology, it is necessary to be very careful with the research methodology employed, choosing a hypothetico-deductive approach that allow us to subsequently evaluate our progresses.

Several authors have highlighted the application of "Design Science Methodology" [276, 281] as an appropriate methodological framework for research into software engineering, but maintaining a great degree of flexibility and innovation required due to the multidisciplinary research in the topic addressed here. Other reason is the fact that research in this area is "oriented towards solutions". In contrast to more observational research "oriented towards problems", in which the objectives focus on explaining and/or understanding a part of the reality, in software engineering a solution is sought for a given problem. This fact may imply a part of observational research (which explains a part of the reality) but it adds a component of definition, design and/or creation of a solution to that problem.

Many of the models of "Design Science Methodology" are based on the fact that a research project begins from one perspective, generally oriented towards a solution. Nevertheless, the research presented here being an example of this, a research project can originate from a great variety of perspectives, each one of them beginning in a different phase of research methodology. This fact is particularly relevant and commonplace in multidisciplinary contexts. For example, we can take a more observational perspective oriented towards a problem during the exploratory phase. However, it is common to change to a more solution-oriented perspective once the problem has been identified and defined and we have moved on to the phase of designing a possible solution. Taking this into account, Peffers' model [213], as opposed to other existing models for the application of "Design Science Methodology", makes it possible to apply various perspectives or to vary the perspective over the course of the research, thus adding flexibility to the application of "Design Science Methodology" [267]. Figure 2.1 illustrates the research methodology of "Design Science Methodology" followed throughout this book, based on Peffers' model [213].

Firstly, the phase "Identification of the Problem and Motivation" defines the research problem to be tackled and justify the value and scientific contribution of the solution. In our case, this phase includes the detailing of the research questions, as well as the whole exploratory review to characterize the problem. Thus, the existing techniques and the designing of tools to explore and characterize the problem described in Part II is also part of this phase. Later, the phase "Definition of Objectives of the Solution" establishes the objective of the solution that we want to design, bearing in mind what is feasible as a solution and what is not. The previous decomposition of the main question in secondary questions and their corresponding solutions created and tested in Chaps. 4 and 5 constituted this phase.

The Design and Development phase includes the creation of the artefact-solution. In our case, this phase includes the design and creation of all the software models which make up the framework proposed. This phase corresponds structurally to Part III of the book. To initially trial the design, the Demonstration phase deals with the question of whether the use of the artefact allows one or more instances of the identified problem to be resolved. In our case, we must demonstrate the framework proposed and their strategies helps archaeologists as far as knowledge-generation processes are concerned. The demonstration is detailed in the Chap. 11.

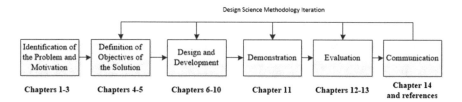

Fig. 2.1 Design science methodology phases and corresponding chapters in this book

Once we have a solution, the Evaluation phase observe and measure the degree to which the artefact (our framework) resolves the problem. This consists of comparing the objective of the solution with the observed results of the use of the artefact in an empirical way. This phase requires, therefore, knowledge of the relevant metrics and techniques of empirical analysis. In our case, an empirical validation has been carried out with the chosen case study, building a software prototype and evaluating it with archaeologists. All of this is dealt with in the Chap. 12 and is discussed in Part V.

Finally, Communication phase includes a dissemination of the results, to communicate the problem and its importance, the artefact-solution, its usefulness and novelty, the precision of its design and its effectiveness for the target users/specialists. Most intermediate results explained in the book are also referenced in the bibliography. In addition, Chap. 14 explains the lines open for future research and their possible implications.

It should be noted that Peffers [213] establishes these phases in an iterative cycle. Thus, it is possible to complete a full cycle for all the phases, for example for each secondary research question which we have posed or for each sub-problem identified in a research project. These iterative cycles have occurred throughout this research, as shall be seen over the course of the book. However, here, a lineal narrative of the methodology has been maintained since this chapter should give a generic overview of the research methodology carried out. Next part reviews existing works on software knowledge generation and what kind of archaeological particularities we should deal with when we want to apply it to large archaeological data sets.

Part II
Software Knowledge Generation
Foundations

Chapter 3
Existing Techniques and Tools

Introduction

Dealing with abstraction processes that takes part in the human mind is not an easy task. As other kind of research areas, knowledge generation processes have been studies from many different perspectives and disciplines with interesting contributions in recent decades. However, only a reduced group of them are specially oriented to the assistance that a software system can provide to humans in generating new knowledge, and even less of them treat this problem as a central part of a co-disciplinary research between software engineering and the knowledge area affected, as is our objective with the archaeological domain. However, this chapter tries to aggregate the ensemble of existing methods, techniques and tools took as a basis in the book, and some of them used later, to explore the problem of software assistance to the knowledge generation process within archaeology in a wide viewpoint.

Thus, there are different types of necessities and works to review as important parts of a software assistance in the generation of knowledge in any domain, though here they are applied to archaeology:

- Conceptual necessities: the generation of knowledge should have solid conceptual bases which allow the issue of software assistance to be addressed. It is necessary to know the conceptual particularities of the field in question. In order to do this, we need tools which analyze the structure of the concepts normally handled in the field and which detect, embody and support their intrinsic characteristics.
- Process necessities: it is necessary to know how knowledge is generated in archaeology on a procedural level and in which points software can assist to this process. In addition, a review shall be carried out of what processes are currently assisted by software in the generation of knowledge in other fields and disciplines, with the aim of finding points of connection with knowledge generation in archaeology.

© Springer International Publishing AG 2018
P. Martin-Rodilla, *Digging into Software Knowledge Generation in Cultural Heritage*, Modeling and Optimization in Science and Technologies 11,
https://doi.org/10.1007/978-3-319-69188-6_3

- Interaction necessities: software assistance to the generation of knowledge is conceived providing information visualization techniques, which are cognitively adapted to archaeologists. It is necessary, therefore, to have methods, techniques and tools at our disposal which allow us to know which mechanisms of presentation and interaction of information are most appropriate for analyzing large archaeological data sets. This will allow us to decide on and/or adapt the visualization techniques to be employed in software assistance.

This chapter is structured according to this typology of necessities. Each section details the methods, techniques and tools which are to be taken as the basis for the exploration of the problem are identified via a review of the corresponding bibliography.

Tools for Detecting Conceptual Necessities

Conceptual Modeling as Knowledge Representation Technique

There is no knowledge generated if we don't have a conceptual structure of data and information about our domain, that is, what are the most important concepts we are going to deal with, the data attributes and possible values and their relationships. This net of concepts is necessary to support any human reasoning (and subsequently any software assistance that we want to create). Thus, a formal conceptualization of each heritage problem, research question and their associated data sets should be our main concern in order to be able to assist to the knowledge generation process in archaeology.

As we exemplified in the archaeological case for illustration in Chap. 1, archaeological projects and their produced data sets often deals with different nature of concepts, not only heritage material referred, but also immaterial, temporal, geographical or more abstract concepts, which are necessary in order to correctly referred the complex nature of an archaeological case. Thus, the following review of conceptual methods and tools covers most of the conceptualization works in Cultural Heritage as a more general discipline, but focusing on how these methods can reflect and represent the archaeological domain.

The formal conceptualization of information handled in the field of Cultural Heritage has been studied by several authors, all with different proposals and different degrees of application and success. Among the most complete proposals is that of the CIDOC-CRM model (CIDOC Conceptual Reference Model) [69, 70], an international standard (ISO 21127) especially designed for "the knowledge of museums" [69]. Although CIDOC CRM was originally aimed at representing the knowledge which experts possessed regarding collections in museums, it has been extended to other areas of Cultural Heritage and is able to absorb much wider types of information. Even so, it is still a model in which material entities are of great importance and non-expert points of view are not easily expressed [106]. Other models which are in existence have other aims, such as the modeling of excavation processes and the archaeological analysis of CIDOC CRM-EH [30] (an extension of the previous model); the composition of a detailed collection of terms relating to heritage in the case of the PHA Thesaurus (Tesauro de Patrimonio Histórico Andaluz) [5]; or support for the interoperability of spatial data regarding protected heritage sites in relation to the Directive 2007/2/EC of the European Parliament establishing an Infrastructure for Spatial Information in the European Community (INSPIRE) [84].

All of these proposals make an attempt to respond to one common characteristic: the existence of a tension between normalization and personalization, between the necessity to establish a common model and the necessity to take into account the fact that each project and effort has its own peculiarities and, therefore, cannot be the object of an inflexible rule [104, 105]. We can find, therefore, both broad and deep models. That is to say, models whose objective is to describe a broad scope (that of a whole discipline, or even a whole world) but which, at the same time, need to specify all the details. This type of model is often seen as being too prescriptive and inflexible, as it leaves little space for the peculiarities of each individual project. On the other hand, there are models which have a more reduced and more superficial scope. In other words, they occupy a limited range (for example, a model of an organization, or even of one project in particular) and avoid going into detailed descriptions. These models can be easily adopted, as it is probable that they are easier to use and more useful in specific cases. However, they are not so useful when it comes to guaranteeing conceptual and technical inter-operability with other models. There is a wide range of possibilities between these two extremes. CIDOC-CRM, for example, has a broad focus in its scope and a moderate degree of depth. CIDOC CRM-EH, on the other hand, is much narrower and much deeper.

However, models whose scope is narrow and not very deep are rarely useful as they provide little added value and those with a more general and deep scope are hardly useable as they are too prescriptive. Viable combinations in terms of scope and level of detail are: (1) models with a general scope but with little detail, or (2) models which are narrower in scope but which have a deep degree of detail. The former are known as models of abstract reference, whereas the latter are called particular or specific models.

With the aim of resolving the tension described above, a semi-formal focus has been presented over the last few years which allows for the creation of models for Cultural Heritage taking this problem into account in an explicit way. This approach, known as CHARM (Cultural Heritage Abstract Reference Model) [99], is a semi-formal representation of Cultural Heritage in the form of an abstract model of reference. In other words, it is a model with a general scope which is not very deep as far as its level of detail is concerned. Therefore, CHARM claims to cover the greatest possible quantity of social and cultural phenomena which are known as Cultural Heritage, albeit at a high level of abstraction. In contrast with other models, such as CIDOC-CRM, CHARM is much broader, due to the fact that it is not only centered on one specific sub-domain (as is the case of CIDOC-CRM regarding museum collections) but on Cultural Heritage in general. In addition, due to its high level of abstraction, it is much less deep: CHARM was designed under the premise that it is necessary to apply extension mechanisms in order for it to be used, whereas other existing models, such as CIDOC-CRM, attempt to be a complete and final solution to be applied directly.

CHARM is expressed in ConML [129] a conceptual modeling language which broadens the conventional focus aimed at objects with characteristics such as temporality and subjectivity modeling, aspects which are especially relevant in areas such as Cultural Heritage [105, 128]. ConML also provides the management and extension mechanisms which are necessary for enabling the extension of CHARM to particular models.

Due to the advantages of CHARM's approach and the appropriateness of ConML to the field of Cultural Heritage [97], CHARM and ConML constitute the conceptual tools employed in this study with the aim of carrying out a structural analysis of the archaeological concepts commonly handled and detecting, reflecting and supporting its intrinsic characteristics.

Basing ourselves on CHARM as an abstract model of reference and using ConML as a modeling language, it is possible to create models which represent realities of differing natures and heritage topics (and especially for our interests, of archaeology cases) but, at the same time, structurally maintaining the issue to be dealt with on a general level. This allows us to create extensions for CHARM for each problem or particular project without losing our structural and semantic reference points of what data is like in the archaeological cases, what areas it covers and what particularities we should take into account. This makes CHARM (and ConML) the most appropriate proposal to be employed here as a model of conceptual and thematic reference as far as heritage is concerned. CHARM specification, as well as the adoption, extension and use which are made of CHARM in this book shall be detailed in Chap. 7.

Tools for Detecting Process Necessities

Existing works on how researchers identify, characterize and assist cognitive processes through software will allow us to know how to identify those processes to assist in archaeology and how to treat them in the software.

Methods for Extracting and Characterizing Cognitive Processes in Knowledge Generation

Extracting Cognitive Information from Textual Sources: Software Strategies

There are different approaches in identifying and extracting which cognitive processes are carried out when we generate knowledge in a discipline: active observation of processes, interviews and questionnaires, analysis of the products resultant in research, methodological analysis and evaluations, etc. Of all of them, we have wanted to go to the primary sources where archaeological knowledge generated is deposited in order to reconstruct backwards what cognitive processes have brought us to it, that is, to reconstruct the "past" (something closely related to the archaeological discipline itself) of the research, from the earliest raw data until archaeologists reach more general and abstract conclusions about the case study they addressed.

In order to do this, and taking into account that in the archaeological practice most of the primary sources are textual sources (reports, fieldwork notes, etc.), we have to deal with a knowledge consolidated in the form of text. Thus, a literature review was carried out of existing studies from a software engineering perspective on the analysis of textual sources and their use as a basis for extracting information regarding how that consolidated knowledge was generated. How is information from non-structured textual sources analyzed and extracted?

The majority of information used as an input in software engineering is originally produced in non-structured formats, such as verbal communication or descriptive documents written in freestyle. We can cite for example documents specifying requirements or documents produced to translate contents into different languages. The non-structured form of these products emerges naturally from the way the participants involved in software engineering processes communicate. However, this situation impedes the rapid analysis of the information, due to the fact that the semantics which are implicit in the textual sources can only be understood by humans and not by way of a highly automated process of information extraction. What is more, non-structured information requires great human effort to be restructured and characterized before being able to be processed to any degree of automation, from the creation of an ad hoc database for the treatment and storage of data, via statistical analysis, to its processing by other semantically similar systems.

Due to this fact, the need for a better conceptualization and structuring of textual information in software engineering has been detected in order, for example, to achieve a higher degree of extraction and fulfilment of the requirements, a significant integration of prior data into new software systems and the generation of appropriate tests for the application of data mining (TDM) [26, 191]. Years ago, Rolland [236] identified four types of strategies for dealing with the relationship between textual sources and conceptual modeling: (1) supporting the generation of models from NL input texts, (2) supporting model paraphrasing, (3) helping in the general understanding of NL input texts by way of modeling and (4) improving NL texts quality. Our work is related to the first strategy as the objective is to extract information existing in texts belonging to the field of archaeology. This information deals with how knowledge is generated within the discipline, in order to be able to characterize and, later, assist it. The extraction of information should generate software models from the text in such a way that we can later deal with this extracted information.

There are two main approaches which can cover the need we have detected: an approach based on information retrieval and an approach based on an ad hoc modeling of the field in question.

As far as the first approach is concerned, specialists in information retrieval have developed a large corpus of approaches in the analysis of textual information as an automatic process, via the use of heuristic and probabilistic approaches and even by the use of semantics [22]. The heuristic and probabilistic approaches focus their attention on the extraction of information from textual sources on a quantitative level [47, 116]. For example, they may extract counts of the number of instances in a specific text by looking at frequency indicators of terms or by implementing mechanisms of automatic indexing. In this way, these approaches allow for quantitative information to be obtained but, in most cases, they do not extract information relating to the semantic relations present between elements in the text. Approaches from within the field of information retrieval, but of a more semantic kind, allow for the analysis of textual sources based on thesauri or thematic maps [134, 208], enabling the extraction of semantic information regarding the under-lying structure of the information in a particular text and the semantic relations existing among its elements. It is possible, for example, to detect relations of equivalence or hierarchy between elements in a text. These semantic approaches have been satisfactorily applied, for example, in order to find common lexemes in a word family, with the aim of analyzing these words as a group [260, 261]. However, the semantic relations which these techniques can extract are not as strong as those obtained via other techniques which have a significant linguistic focus, such as those based on discourse analysis. The latter, as well as being able to extract equivalences and hierarchical relationships, can also detect and extract relationships of causality and exemplifications present in the text. In addition to the limitations mentioned regarding both approaches (heuristic and/or probabilistic and semantic), all approaches originating from fields based on information retrieval need to work at a high level of abstraction as far as techniques are concerned in order to be truly independent solutions from the field of application and to be able, therefore, to extract information from documents written in freestyle. This fact does not allow work to be carried out at a level which maintains the semantics implicit in complex narratives, as is the case of archaeological documents, an aspect which is taken into account by more linguistic approaches.

Continuing in the realm of information retrieval, in a later paper, Rolland [233] classifies the current techniques of the automatic generation of conceptual models from textual sources taking into account the following characteristics of the generated model: on the one hand, static or dynamic models; on the other hand, models based on rules or on ontologies.

As far as static aspects of the conceptual models are concerned, tools exist which enable us to discover and generate object models or class models concerning the specifications of textual requirements [113, 146, 186, 192]. As for the dynamic aspects, Rolland highlights studies which extract cases of use and scenarios of texts [154, 182, 234, 241]. As for whether the models are based on rules or on ontologies, there are approaches which attempt to discover business rules in business process models [119].

Even if all of these studies serve as a basis to enable us to approach our aim of structuring and extracting semantic information from textual sources, none of them take into account structural elements (phrases, clauses, etc.) of the text itself within the conceptual models they generate. This does not allow the model associated to the texts of origin to be maintained, an aspect which is of necessity in archaeology. What is more, the approaches listed here center their attention on automatic processes, which generally do not form part of a complete software engineering methodology but rather they are carried out separately, and generally prior to the execution of the methodology (for example to extract requisites from texts before beginning with a specific methodology). We are seeking an approach which allows us to integrate this process of structuring and extraction of semantic information from textual sources within the complete methodology of software engineering, which is applied at all times. Therefore, the generation of these models is generally carried out by software engineers, who later evaluate the resulting models in collaboration with specialists in the field in question. In an attempt to maintain our aim of carrying out a genuinely interdisciplinary study which does not submit any of the disciplines involved to an auxiliary or user role, we are seeking a solution which allows specialists in archaeology to create their own models, thus adequately structuring the information and extracting from the text the most valuable information in order to generate knowledge in their field.

With regard to the second approach, the solutions based on ad hoc modeling of the field of application for the extraction of information originating from textual sources have undergone a significant increase in recent years, especially those related to the application of textual analysis to fields with a high degree of necessity for the structuring of information, such as the field of Biomedicine [51, 144]. These solutions have been applied satisfactorily to give structure to non-structured texts. However, they only work within a well-defined context within a specific field of application. Once this limitation is assumed, it is possible to create an extremely precise conceptual model which captures the semantic relations which appear among elements within a specific context. However, this resulting conceptual model needs to be created ad hoc for each application: a new textual analysis implies the creation of a new conceptual model. For these reasons, it is not possible to achieve a high degree of standardization as far as the conceptualization and structuring of information is concerned.

The limitations which appear in the structuring and extraction of semantic relations in texts written in freestyle which are used as a source in software engineering motivate our proposal for a methodology integrating techniques of discourse analysis in software engineering (and presented in next Chap. 4), enabling the representation of the elements of textual information by way of a particular language in a structured way and describing the semantic relations which exist among them. The methodological proposal, as well as the language created, is based on the ISO standard *ISO/IEC 24744 Software Engineering—Metamodel for Development Methodologies* [140], with the aim of facilitating the connection between the language and the methodological proposal and its integration into any other methodology of software development expressed in the standard.

The following section explains the work relating to the formalization and analysis of textual discourse, detailing its potential for structuring and extracting semantic information compared with the existing approaches listed above. All these existing works serve us as a basis for the methodological proposal presented in Chap. 4.

Extracting Cognitive Information from Textual Sources: Discourse Strategies

As has been mentioned before, there are some needs relate to the improvement in structuring and extraction of semantic relations presented in textual sources if we want to analyze and extract cognitive information from them. In other words, although the semantics present in a text are self-contained within it, any software engineering process needs to make these semantics much more explicit and precise, avoiding ambiguity. Discourse analysis techniques are able to give structure to these textual sources, which are in a non-structured format, and to extract the semantic relations present within them.

The term "discourse analysis" was originally coined by Harris [115], who defined it as "a method for the analysis of the connected speech or writing for continuing descriptive linguistics beyond the limit of a single sentence at a time and for correlating culture and language" [115]. All of the techniques which later arose from this first definition constitute a broad field in Linguistics, above all due to the need to discover meaning in terms of the narrative elements present in discourse and focusing analysis on the organization of the language, above or below the levels identified as sentence or paragraph. The choice of one technique of discourse analysis or another depends on numerous factors, such as the sphere of application or the objective of the analysis: the structuring of the discourse in phrases or syntagms, the identification of functions of particular elements within the discourse and the representation of texts, etc. [107].

Currently, discourse analysis techniques are not only used in Linguistics, but are also satisfactorily applied, with different degrees of automation, to a significant number of fields in different contexts, albeit with the common necessity to study oral communications or textual sources. Good examples of their application can be found in Biomedicine [158] and the extraction of analytical information in legal texts [187] with applications focused on identifying relations of consequence among elements in the text in order to detect possible consequences of a certain legal action. In all these cases, the discourse analysis carried out allows the elements present in a text to be identified. These elements provide information about the structure of the narrative and the reasoning and intention of the author. In addition, these analyses help us to understand the most commonly used cognitive processes in the field in question.

The majority of approaches in discourse analysis with empirical aims attempt to characterize the semantic relations intrinsic to the discourse, joining elements of the discourse (typically phrases, although it is possible to work on a more detailed level, such as syntagms, or on a higher level of abstraction, such as complete paragraphs). These approaches with empirical aims follow three courses as far as the semantic relations to be extracted are concerned [251]: There are studies focused on creating a corpus or typifying coherence relationships, such as the RST corpus [42], the SFU Review Corpus [258], CAuLD (*Construction Automatique de representations Logique du Discours*) [19, 20] and the Penn Discourse Treebank [223]. There are also studies which have a clear psycholinguistic leaning, presenting empirical annotation experiments in which the subjects are presented with real data to be annotated or examples of textual fragments, with the aim of experimentally identifying which characteristics are important for the semantic relations in the text. Furthermore, there are studies which make use of an extremely extensive corpus (*Very Large Corpora*) in order to find, automatically or semi-automatically, characteristics of types and to identify different coherence relations by way of the identification of explicit connectors, such as Marcu [42], Lapata and Lascarides [160], Reitter [231], Chang and Choi [58].

In summary, all of the studies listed above illustrate the demonstrated potential of techniques of discourse analysis in identifying semantic relations and structuring textual sources on the level of discourse elements. It can be considered, therefore, that discourse analysis is a flexible and appropriate basis for structuring and extracting semantic relations from texts written in freestyle, including those which are used as a source in any software engineering process.

Although there are specific applications which allow us to analyze a set of documents, thus avoiding an ad hoc discourse analysis for each document, these are not formal or abstract enough to be integrated into a complete Software Engineering methodology. They are not formal enough because they are never expressed as a metamodel or a similar unequivocal mechanism and they are not abstract enough because, as has already been mentioned, they often require ad hoc work. For these reasons, we believe that a complete methodological proposal is necessary, along with a language with a general purpose which allows for the application of discourse analysis to heterogeneous areas as the use of the proposed language must present well-defined methodological components. Only by using a completely defined methodology can discourse analysis techniques work as part of an integral methodology for software engineering.

In order to achieve this, the decision was taken to follow Hobbs' [120] approach to discourse analysis, due to the fact that (1) it allows for the characterization of semantic relations based on cognitive processes among elements of the discourse,

(2) it presents a well-defined method of application which allows us to express the methodology of the discourse analysis in question as a formula and (3) it has been previously applied in software development [180, 220], although it has not been formalized either in terms of language or in methodological terms for this purpose. In addition, Hobbs' method has been used with narrative texts of different types and has been expressed in different languages [153], which guarantees us a certain degree of universality in these aspects.

Hobbs' work is based on the formal identification of relationships between elements of discourse, so-called coherence relations. Coherence relations in discourse, similar to those of contrast, generalizations or causal relations, contain information which is essential to understanding how the elements of a fragment of text and the underlying ideas referring to any domain are related. Any computational system which attempts to understand or generate information beyond the level of a simple phrase must deal with this kind of information. Hobbs' work [120] constitutes, as has previously been mentioned, one of the most important theoretical approaches in existence with proven applications. Our methodological proposal for extracting cognitive processes from textual sources based on Hobbs' work [120] is detailed in Chap. 4.

Existing Cognitive Processes Characterizations

Apart from different approaches based on extracting cognitive information from textual sources to analyze the knowledge consolidated, there are other set of research working in reverse (i.e. from top to bottom), trying to firstly obtain a cognitive processes characterization and then test it in the real knowledge generation cases. Next sections review these approaches, giving the reader a broad overview of the topic.

Theoretical Models for Knowledge Generation

Some theoretical models regarding knowledge generation exist which constitute one of the main corpus of this research. All of the existing models follow a hierarchical structure based on layers: Cleveland [49] establishes a model with four layers: Facts and Ideas, Information, Knowledge and Wisdom. Cleveland's model lays the foundation for a theory of human understanding. The intermediate processes between layers are not detailed in this study. Later, Ackoff [7] went a step further with five layers: Data, Information, Knowledge, Understanding and Wisdom. This model of knowledge generation has been used for years as a point of reference in Psychology and Cognitive Studies. Taking Ackoff's model as a reference point, other authors have proposed their own models, always similar in structure but different in terms of the semantics of the cognitive processes involved. Carpenter and Cannady [45], for example, take other characterizations of the intermediate process between layers as a basis [27] and incorporate feedback

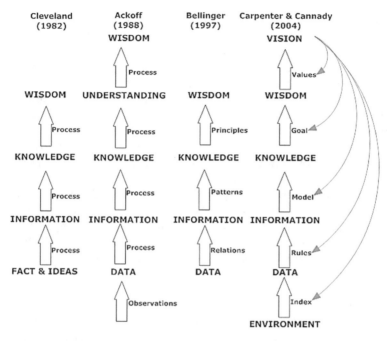

Fig. 3.1 Existing theoretical models to characterize cognitive processes in knowledge generation. Image based on Scott A. Carpenter [44]

between layers. They propose a model with six layers: Environment, Data, Information, Knowledge, Wisdom and Vision.

In summary, the existing models for knowledge generation coincide in determining a model based on layers or levels, in which each transition from one layer to the next is brought about by humans carrying out cognitive processes, whose names and spheres differ according to the author. A summary of the models describe can be seen in Fig. 3.1.

The Role of Cognitive Processes in Software Engineering

The cognitive aspect is an area which is currently gaining in interest and importance both within Industry and in the research area of software engineering. Over the last few years, both the human and the specifically cognitive aspects within software engineering and information systems engineering have received growing attention in the literature and in relevant conferences, proving that these aspects are just as important as the technical ones, which have traditionally been dealt with in greater detail. A good example of this is the upsurge in scientific forums which treat cognitive processes as a central theme in software engineering, such as the

COGNISE workshop [50] and the numerous papers which, from different perspectives, consider that cognitive aspects in the field play an essential role.

Although previous studies exist which deal with cognitive processes in software engineering (always from the perspective of the developer, designer or creator of the software) [36], the decision was taken to carry out this review of the most recent studies carried out regarding cognitive processes in the discipline, due to the fact that it was not until recent times that these cognitive processes were treated as an instrument to assist to the process of knowledge generation.

On the one hand, there are recent studies which deal with cognitive processes within the process of software development itself, providing answers to questions such as: What cognitive processes do the agents involved perform in the development of software? How do they perform them? These studies focus on cognitive processes which are carried out during the tasks of modeling [216, 217], design and programming of software artefacts [112, 265], in addition to dealing with issues of collaborative work/teamwork in these environments. Furthermore, similar studies carry out research on what cognitive processes are performed by the users who receive the software products generated during the process of software development [277].

On the other hand, studies [209, 268] can be found which are more oriented towards the study and in-depth description of specific cognitive processes. These studies, in spite of being situated within software engineering contexts, have a more theoretical and independent approach from the field and provide answers to questions such as: What is the role of the cognitive process of classification in software engineering? Is this cognitive process of general classification for any discipline or does it present particularities in the field of software engineering?

Due to the growth of both areas of study, the task of carrying out a review has become a difficult one and lies beyond the scope of our purpose. However, a review of the area which concerns us has been carried out: our efforts have been focused on identifying studies which describe cognitive processes in contexts of software assistance to knowledge generation independently of the techniques used for assistance.

The majority of assistance software systems which were identified have the aim of assisting the user to carry out physical processes (involving tasks with the software itself), for example, in industrial contexts with a high degree of automation, or in medicine, etc. There are other systems which are closer to assistance to the generation of knowledge, mainly in arithmetical studies and decision-making assistance, the analysis of textual argumentation and information visualization.

As far as the former are concerned, these systems focus on the access and conceptualization of the data which is being handled [288], and very few of them incorporate models of cognitive processes. The incorporation of these models is defined in some studies in applied psychology to studies in mathematical analysis, such as those revised by Ashcraft [18]. The characterizations shown in this type of study cover reasoning of a purely arithmetical kind, with generally very high levels of abstraction in the definition of the cognitive processes and quite underdeveloped formalizations. For example, characterizations of cognitive processes can be found referring to Association, Confidence Criterion and Search Length [39] or in terms of Categorization/Discrimination [25]. This same problem presents more current characterizations based on those previously mentioned, with spheres of application focused on decision-making. These characterizations prove useful for the context of application in empirical studies with human beings or when working on the definition of systems with a high degree of abstraction, though not in contexts of software assistance, in which a specific characterization of certain cognitive processes to be assisted is needed. Later, the importance of certain specific cognitive processes identified in these studies, such as classification and discrimination, has indeed given rise to specific conceptualizations [209] and applications for those processes, albeit without comprising a wider characterization of cognitive processes in software assistance.

In the case of studies which deal with cognitive treatment from textual sources, it must be highlighted (as has already been explained in detail in previous sections of this chapter) that some approaches to discourse analysis deal specifically with cognitive processes of argumentation, such as their classification in the form of Hobbs' coherence relations [120], explained above.

In the case of systems relating to information visualization, there are studies which do imply cognitive processes in an explicit way and which relate them with the other elements of a specific framework. They are designed independently from the field of application and are only focused on cognitive processes which play a relevant role in information visualization. Good examples of this are Zhou [291], Amar [12] and Yi [288]. All of these characterizations show low or average levels of abstraction and are used successfully in the area of information visualization. However, the cognitive processes which they characterize are only focused on visualization tasks. Therefore, the direct use of one of these characterizations for our objective of software assistance to knowledge generation would leave out any other assistance technique which we may wish to include at a future moment (such as the techniques relating to knowledge extraction which were outlined in Part I of this book).

Within the sphere of information visualization, Chen's work [61] is worthy of note due to the fact that, although it does not present a specific characterization of cognitive processes, it includes a full model for knowledge generation via software assisted visualization. This model has been applied in a satisfactory way, mainly in the field of Biomedicine, with its definition being independent of the field of application. Chen proposes a model which captures the actions of the user, establishing how he/she is generating knowledge. In the next step, the system adapts its behavior to the user and offers him/her assistance by way of specific visualization tools. The authors of this model make reference to the DIKW hierarchy as the framework for their model, thereby situating it within a sphere of application which is appropriate for our purpose. In spite of the fact that this model could be of use to us as it is, we consider it necessary to include a specific characterization of cognitive processes within the model itself. In addition, our objective is to characterize a framework for archaeology, not only with software assistance by way of visualization but also taking into account possible future extensions through the use of knowledge extraction techniques. Bearing this in mind, Chen's model serves us as a point of reference which will be looked into fully in later sections.

In conclusion, the majority of studies focus on cognitive processes related to the development of the software itself and not on studying them from the point of view of software assistance, in spite of the importance which cognitive aspects have acquired in software engineering in recent years. The studies which have most in common with our aim of providing assistance either work at a high level of abstraction with a low degree of formalization (studies in cognitive psychology, arithmetic and decision-making), thus impeding the inclusion of these characterizations as a part of a software model, or produce characterizations for a specific sub-domain (for example, information visualization or the analysis of textual argumentation).

Characterizations of Cognitive Processes in Archaeology and Related Heritage Domains

Just as in software engineering contexts, the characterization of the cognitive processes involved in archaeology has gained relevance in recent decades with studies such as those by Stockinger [256], Gardin [90] and, more recently Doerr, Kritsotaki and Boutsika [71]. Another important set of studies which work on a greater level of abstraction are those related to the identification of methodological primitives in Cultural Heritage studies, such as those by Unsworth [266], Palmer [204], Blanke [31] and Bernadou [28] which focus more on defining cognitive processes as tasks within the characterization of the process of knowledge generation.

However, the level of formalization of these studies is not high enough to enable us to verify our initial hypothesis and to obtain results which can be used directly in software assistance systems in knowledge generation. Table 1.1 provides a synthetic and comparative vision of the specific characterizations of the cognitive processes which have been studied. The comparison has been made based on four criteria:

- The level of abstraction: There are four possible values (HH = EXTREMELY HIGH, H = HIGH, M = AVERAGE AND L = LOW), according to whether the characterization concerns very specific cognitive processes (involving specific tasks) or more general ones.
- The disciplinary sphere of origin: This indicates in which discipline the definition of the characterization is set, using the following abbreviations: (PSY = PSYCHOLOGY, SE-DSS = SOFTWARE ENGINEERING-DECISION SUPPORT SYSTEMS, INFOVIS = INFORMATION VISUALISATION, LING = LINGUISTICS, CH-METHOD = CULTURAL HERITAGE-METHODOLOGY STUDIES, AR = ARCHAEOLOGY).
- The degree of formalization: There are four possible values (HH = EXTREMELY HIGH, H = HIGH, M = AVERAGE AND L = LOW).

Table 3.1 Main characterizations of the cognitive processes studied

Source	Characterization	Level of abstraction	Disciplinary sphere	Degree of formalization
Groen and Prakman [110]	Direct memory retrieval, back-up counting	HH	PSY	M
Ashcraft [18]	Retrieval, spreading activation	HH	PSY	M
Campbell [39]	Association, confidence criterion, search length	M	PSY	M
Beale [25]	Categorization, discrimination...	H	PSY	L
Poldrack [222]	Attention, language, memory, music, reasoning, soma, space, time	H	PSY	HH
Schwenk [245]	Goal formulation, problem identification, alternative generation, evaluation/selection	H	SE-DSS	H
Chen and Lee [60]	Case memory, cognitive mapping, scenario building	HH	SE-DSS	M
Zhou and Fenier [291]	Associate, background, Categorize, cluster, compare, correlate, distinguish, emphasize, generalize, identify, locate, rank, reveal, switch, encode	L	INFOVIS	H
Amar [12]	Retrieve value, filter, compute derived value, find extremum, sort, determine range, characterize distribution, find anomalies, cluster, correlate	M	INFOVIS	H
Yi [288]	Select, explore, reconfigure, encode, abstract, filter, connect	M	INFOVIS	H
Hobbs [120]	10 coherence relations: causal, generalization, exemplification...	M	LING	H
Unsworth [266]	Discovering, comparing, selecting, linking, sampling, referring, illustrating	H	CH-METHOD	M
Palmer 2009 [204]	Browsing, collecting, rereading, assembling, consulting, notetaking	H	CH-METHOD	M
Blanke [31]	Discover, compare, collect, deliver, collaborate	H	CH-METHOD	M

(continued)

Table 3.1 (continued)

Source	Characterization	Level of abstraction	Disciplinary sphere	Degree of formalization
Bernadou [28]	Berry-picking, chaining, combining, annotation, thematic organization, translation, and database development	H	CH-METHOD	M
Stockinger [256]	Logic propositions: analogies, if-then structures and conceptual inference (as a higher level of abstraction)	H	AR	L
Gardin [90]	Logic propositions: analogies and if-then structures	H	AR	L
Doerr [71]	Factual argumentation, inference making, belief adoption	HH	AR	M

The codes used for their classification are explained in the text preceding the table

A high or extremely high degree of formalization corresponds to characterizations in which we have not only identified the cognitive processes but have also defined and structurally described them, for example in the form of mathematical parameters, algorithms, rules of association, etc. Average or low degrees of formalization correspond to characterizations in which we have only narratively defined the primitives of the characterization.

Most characterizations and techniques reviewed here to extract cognitive information from textual sources are taken as a basis to propose the characterization of archaeological cognitive processes that we are going to treat during the next chapters. The proposal will be detailed in Chap. 4.

Tools for Detecting Interaction and Presentation Necessities

Determining the Suitability of Interaction and Software Presentation Mechanisms

Finally, this extensive review would not be complete without analyzing how represent and validate interaction and data presentation strategies in large data sets, in order to serve as a reference for our software assistance solution for archaeological domain.

Many approaches have been adopted to determine whether a mechanism of interaction and/or the presentation of information via software is appropriate or not for its purpose, with studies on heterogeneity, the methodology of application and links to other disciplines, such as Psychology or Sociology. Due to the broad scope of the topic, only those techniques and methods employed in the current literature which allow us to determine the suitability of existing information visualization techniques to specific aims have been selected. In this area, validation studies also abound [13, 193], grouped into frameworks [21], although they coincide in the empirical approach to these validations. In 2009, Munzner [193] carried out an in-depth study regarding techniques with an empirical focus which are used in the validation of information visualization techniques. She analyzed a set of groups: algorithm complexity analysis, field study with target user population, implementation performance (speed, memory), informal usability study, laboratory user study, qualitative discussion of result pictures, quantitative metrics, requirements justification from task analysis, user anecdotes (insights found), user community size (adoption) and visual encoding justification from theoretical principles.

Following these works, it is evident that information visualization techniques and their adaptation to specific domains and validation inside them requires a deep research on suitability working with real data sets. In our case, it is necessary to determine the suitability of existing mechanisms in information visualization techniques for carrying out cognitive tasks in archaeology, so we considered it necessary to design a series of empirical studies using techniques which involve not only the appearance but also the cognitive processes which they support. In addition, we believe that a mixed analysis, which allows for the extraction of quantitative (and thereby scientifically reproducible and verifiable) and also qualitative (allowing archaeologists to express themselves in a less structured way) information, will work better in the context which concerns us. For these two reasons, studies with users in the laboratory, though partly defined in terms of quantitative experimentation in software engineering [155], constitute the most appropriate formal framework for determining the suitability of certain existing interaction and software presentation mechanisms to be validated in archaeology. The entire experimentation design of these previous experiments and their results, as an initial idea of what kind of information visualization techniques we need to work with, are detailed in Chap. 5.

Chapter 4
Dealing with Archaeological Particularities

Introduction

Now that we have contextualized and motivated why it is important a co-research approach to assist via software knowledge generation processes, we are ready to focus on the archaeological domain and their particularities. As we have showed in previous chapters, archaeological data sets (and in general the conceptualization of the archaeological entities and processes) present some characteristics that, without represent unique situation because most of them are common in other humanistic disciplines, their treatment is essential for an effective approach to the topic. Most representative characteristics could be list as:

- Underlined conceptual importance of temporal and geographic dimensions of the archaeological data. In addition, needs for dealing with conceptual vagueness and subjectivity inside them.
- High degree of variability in data sources.
- The data sets (specially the large ones) are commonly design and produce for multidisciplinary teams.
- Preferential use of textual sources, as well as inputs for the creation of data sets (raw data from which the archaeologist generated new knowledge) and outputs (final reports with generated and consolidated knowledge based on the previous data analysis).
- Lack of studies about what cognitive processes are more commonly employed as part of the reasoning handled and applied in archaeology.
- Very recent and limited studies (mostly fragmented ones) about how the visualization of the analyzed data sets affects the archaeological knowledge that finally is generated, which of the existing methods to visualize them presents better results and how to improve the visualization and interaction software for analyze archaeological data.

© Springer International Publishing AG 2018
P. Martin-Rodilla, *Digging into Software Knowledge Generation in Cultural Heritage*, Modeling and Optimization in Science and Technologies 11, https://doi.org/10.1007/978-3-319-69188-6_4

Considering all these particularities, and after the review on existing works (both from software engineering and from a more humanistic vision and cultural and archaeological works), this chapter presents the strategies created throughout this research for dealing with most of these archaeological particularities, representing contributions to improve particular aspects of them. Specifically, the chapter details: (1) a methodological proposal for integrating discourse analysis in software engineering in order to analyze subsequently archaeological discourse and to extract from them relevant information for our software systems, (2) an original characterization of the most common cognitive processes identified in the archaeological textual sources, which allows software assistance to be used in the generation of knowledge, (3) a TAP protocol to validate cognitive processes and (4) TAP recommendations to perform information visualization empirical studies with archaeologists.

These contributions will be subsequently empirically tested (see Chap. 5), and all of them are already a basis for any reader in case he/she wants to apply them directly (autonomously to the framework that the book later presents). However, they will be integrated to create software to assist in the analysis of archaeological data them into the framework (Part III). Next sections specific each contribution one by one.

Archaeological Consolidated Knowledge Particularities

A Methodological Proposal for Integrating Discourse Analysis in Software Engineering

Practices related with the generation of knowledge in archaeology, such as archaeological excavations, interviews, anthropological studies, geo-environmental analyses in the laboratory, architectonic studies and linguistic aspects, produce a great deal of data. The majority of archaeological knowledge generated is contained in reports and monographs in the academic realm and in textual documents such as administrative reports. These documents (both research reports and monographs and administrative reports) are in the form of narrative and are barely structured. What is more, their use is normally limited to the scope of a project or activity, be it academic or administrative. Thus, it is necessary a methodological path for the treatment of this consolidated knowledge, as a source from which information could be extracted on how knowledge is generated in the field, what problems are presented by this generation of knowledge and what possibilities and necessities of software assistance there are.

Firstly, detailed explanation shall be given of the ISO standard *ISO/IEC 24744* [140], the chosen medium for expressing the methodological proposal and the language created as tools in order to structure and extract semantic relations in descriptive freestyle texts. Then, the methodological proposal specification is described, detailing its potential for structuring and extracting semantic information.

ISO/IEC 24744 Standard

In general, a modeling language, whatever its scope or purpose, will be used in a specific methodological environment. That is to say, it will be used by different people playing different roles in tasks and/or processes which use, believe in, modify or reject physical or conceptual artefacts. Due to this fact, it is necessary for this methodological environment to be made clear and taken into account, along with syntactic and semantic aspects, when it comes to creating a new modeling language. Therefore, the language is generally expressed as an instance of a standard metamodel for methodologies. This option facilitates the methodological integration of the language, avoiding inconsistencies [100] and making the context of its use clear.

With the aim of making the methodological context of the modeling language which is created explicit, the ISO standard *ISO/IEC 24744* has been chosen as the metamodel of reference. *ISO/IEC 24744* provides us with the basic conceptual constructs to define a modeling language, as well as integrating it with previously defined processes which involve people, tasks, etc., an ability which is lacking in other metamodels such as OMG's SPEM [101, 201].

The ISO standard *ISO/IEC 24744* consists of several main classes:

- The ModelKind class represents a specific type of model which can be used in a methodology, for example, models of classes, of processes, etc.
- The Model class represents a particular model which is built and/or used within a performative action in the context of a methodology. For example, a model of classes in particular or a specific model of processes.

Each model is of a specific type, and this is reflected in the ISO standard ISO/IEC 24744 via the concept of powertype, which plays a crucial role. The powertype concept was introduced in software engineering by Odell [199] and was applied to metamodeling in next works [102, 117]. In Fig. 4.1, the ModelKind model and classes are shown together as they constitute a power type pattern. That is to say that ModelKind is a powertype of Model. This implies that the instances of ModelKind are also subtypes of Model and, therefore, any Model class in particular which we want to define following ISO/IEC 24744 would be shown by an object (an instance of ModelKind), likewise by a class (a subtype of Model); this situation, in which an object and a class represent the same thing gives rise to a hybrid entity known as a Clabject.

- ModelUnitKind and Model Unit,—see Fig. 4.1—These form another power type pattern, in which ModelUnitKind represents a modeling primitive which

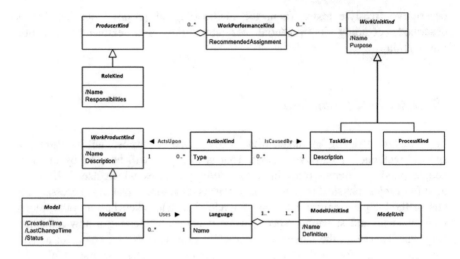

Fig. 4.1 Fragment of metamodel ISO/IEC 24744 with its main classes

may be part of a language, such as "class" or "attribute" in the modeling of classes, or "process" or "task" in process modeling. Model Unit, on the other hand, represents a particular case of ModelKindUnit, for example, a specific class or attribute, or a specific process or task following the analogy with process modeling.

- The Language class represents the language used to express each type of model.
- The methodological integration of a language based on *ISO/IEC 24744* is achieved via the semantics embedded in the metamodel: as can be seen in Fig. 4.1, the ModelKind class is related with Task Kind (via Action Kind). This permits the process which a model uses, creates or modifies to be easily captured through these types of action (creating, reading, modifying or deleting). In turn, Task Kind is related with Producer-Kind (via Work Performance Kind) so that people or tools participating in these actions can express themselves easily [140].

The remainder of the standard's metamodel will be used by way of the powertype mechanism detailed above and following clause 8.1.2 of the standard's specification [140].

Once the fundamental concepts of ISO standard *ISO/IEC 24744* have been reviewed, we will focus on the Language class, whose semantics provide us with the opportunity to create an instance of a modeling language which allows us to handle concepts of discourse analysis and apply them to any software engineering process in which it is necessary to structure and extract semantic information from textual sources. The following section will include a full description of the proposed methodology and the modeling language which has been created.

The Proposed Methodology

With the objective of structuring and extracting semantics from freestyle textual sources, as well as carrying out discourse analysis processes in an integrating way in any defined software engineering methodology, the proposed solution is organized into two parts: (1) a modeling language capable of capturing the structure and semantics implicit in a textual source and (2) a proposal to describe the methodological elements involved in the process of discourse analysis.

The Modeling Language for Discourse Analysis

This section contains a description of the proposed modeling language. Figure 4.2 shows how the most specific classes of the metamodel corresponding to ISO/IEC 24744 have been used to form the basis of the proposed language.

Firstly, the Language class has been instantiated, creating an object *L1*, whose property Name presents the value "Discourse Language". This object represents the language itself.

In addition, as was mentioned in the previous section, the powertype patterns for *ModelKind* and *Model UnitKind* have been used to create instances of the corresponding clabjects. In the case of *ModelKind*, the *Discourse Model* class has been created along with an object *MK1*, which are, respectively, a subtype of Model and an instance of *ModelKind*. Both form a clabject, represented in Fig. 4.2 by a dotted ellipse; this clabject represents the specific type of models which can be created by way of the use of the language *L1*, which we have just defined. This specific type of model is known as a "Discourse Model". Therefore, any model of discourse is expressed in the proposed language for discourse analysis.

Furthermore, the class *DiscourseModelUnit* and an object *MUK1* have been created, which are, respectively, a subtype of *ModelUnit* and an instance of *ModelUnitKind*. Both form a second clabject, which represents the primitives of

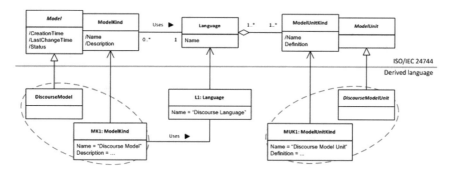

Fig. 4.2 Instances of ISO/IEC 24744 used for the creation of the defined modeling language

modeling units which form part of the language L1 and which can be used to compose models of the "Discourse Model" type.

Given that the final objective of the proposal is to provide structure and extract semantic relations in discourse analysis, the language should contain elements from three different and semantically relevant areas for this aim: elements from the discourse itself (narrative elements), elements from the field in question (what the discourse being analyzed is about) and the coherence relations which may exist between them (the semantic relations defined by Hobbs). In the first place, we have to represent the discursive elements present in the textual source, such as phrases, clauses and their aggregates. This area is represented by the *DiscourseElement* class and all of its specializations, shown at the center of Fig. 4.3.

Secondly, any discourse always refers to a part of the reality and, therefore, the field in question should be taken into account. Following the modeling approach oriented towards objects, this field is structured around entities (objects) and their values, as well as types (classes) and characteristics (attributes). These primitives are not destined to substitute a complete modeling language such as UML [139] or ConML [129] but to work as a mapping point with the field in question in the process of discourse analysis. This area is represented by the DomainElement class and all of its specializations, as shown on the left of Fig. 4.3.

Having specified these two areas in our language (the elements of the discourse and the entities of the reality which it refers to), we will then be able to model which fragment of the discourse refers to which specific entities in the field in question. The "Reference" class acts precisely as a connector between these two realities and,

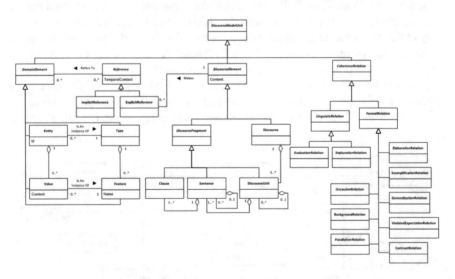

Fig. 4.3 The proposed modeling language. The image shows a complete view of the three areas represented, from left to right: the field described, the associated textual representation and the coherence relations which connect the two previous areas

optionally, may maintain temporal information with the aim of supporting temporal changes in the references throughout the discourse. These references can also be divided into explicit references (*ExplicitReference*), those whose reference to determined entities of the field is made clear in the text being analyzed, and implicit references (*ImplicitReference*), those which, in spite of referring to elements of the field in question, do not make reference to them in the text being analyzed, thus working on an elliptic level. As can be observed in Fig. 4.3, any *Reference* is mapped with the element of the field to which it refers. However, it is a common occurrence that in texts written in freestyle some references are implicit. In these cases, it is usual for the implicit references not to be formalized in the discourse analysis and the semantics related with them to be lost. With this mechanism, the proposed language is able to formalize implicit references and to deal with them in a coherence relation, thus maintaining their semantics.

Finally, a modeling language for discourse analysis must consider a third area, which, as has previously been mentioned, provides the maximum value of this contribution. Following Hobbs' approach, we modelled the different coherence relations which may be present in the discourse via the use of a class to represent each relation. This area is represented by the *CoherenceRelation* class and all of its specializations, as shown on the right of Fig. 4.3.

Figure 4.3 also shows that the coherence relations identified by Hobbs are organized into two categories which are reflected in two abstract classes: linguistic relations and formal relations (*LinguisticRelation* and *FormalRelation*). This classification is not present in Hobbs' work but has been incorporated into our contribution due to the fact that it is useful from the perspective of metamodeling. In this way, we can deal with the degree of detail with which the entities of the field are referred to. On the one hand, formal relations allow us to formally describe which elements of the discourse refer to which entities and the values of these entities in the field with a high degree of detail. On the other hand, in linguistic relations, it has not been possible to reach this degree of formalization. It has proven more difficult to abstractly assign the relations and the elements of the field in question. In spite of the fact that it is possible to do it case by case (in the models of created objects), we have preferred to maintain an unmapped metamodel specification of the proposed language.

Each type of coherence relation makes references to the field in a particular way, which is why each *Coherence Relation* subclass will be associated with the Reference class in a different way. The proposed language incorporates the ten coherence relations identified by Hobbs. The following sections shall explain in detail each coherence relation: the formalization proposed for it, the metamodel expressed in the language of discourse analysis created and some OCL restrictions [278] added in order to guarantee the coherence and structural consistency of each metamodel.

Occasion Relation

According to Hobbs' definition, an "Occasion Relation" is one of coherence between two elements of discourse which describes two events: "the first event sets up the occasion for the second. In both cases we let S_1 be the current clause and S_0 an immediately preceding segment". There are two cases:

A change of state can be inferred from the assertion of S_0, whose final state can be inferred from S_1.

A change of state can be inferred from the assertion of S_1, whose initial state can be inferred from S_0.

According to Hobbs' definition, we describe the "Occasion Relation" with the formula:

$$V_{ijA} \text{ occ } V_{ijB}$$

where each $V_{ij}T$ is a fragment of the discourse which refers to a value V of the characteristic F_j of an entity E_i, in different temporal situations: T_A, named initial and T_B, named final. This means that the entity E_i changes state between moments T_A and T_B, as a result of the modification of the value V of its characteristic F_j. In Fig. 4.4, the proposed language captures the structure and semantics of the relation.

As can be seen in Fig. 4.4, each *OccasionRelation* includes two references which play an initial role and a final role, mapping a fragment of discourse to one or more values. The values mentioned by each one of these fragments of discourse should belong to one common entity. Therefore, we can use the language to express the change in the values over time (given by the different instances of Value), in addition to the associated temporal sequence.

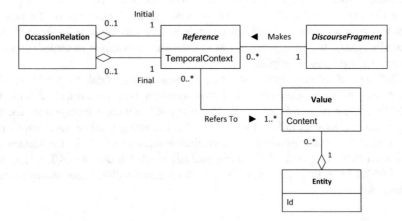

Fig. 4.4 Metamodel for the occasion coherence relation

It should be noted that, both in the Occasion coherence relation and in the others defined by Hobbs [120], specific examples can be found referring to different fields in previous studies forming part of this doctoral research project [171].

Background Relation

According to Hobbs' definition, a "Background Relation" is one of coherence between two elements of discourse in which one element provides contextual information for a second segment. It provides the "geography" (in the words of Hobbs) in which the events of the second series of sessions are carried out. Hobbs clarifies this by establishing a formal definition: "Infer from S_0 a description of a system of entities and relations, and infer from S_1 that some entity is placed or moves against that system as a background".

According to this more formal definition, we describe the "Background" relation with the formula:

$$Ei \, bkg \, Ai$$

where each Ei is a fragment of the discourse which makes reference to an entity Ei. The fragments of discourse play the role of "subjects". Each Ai is a fragment of the discourse which describes contextual information about the entities of Ei, playing the role of "context". Therefore, we need at least a couple of fragments of discourse, an E1 and an A1, in order to identify a "Background" relation, according to Hobbs' definition. However, our formalization allows us to identify more fragments of the discourse which act as context and which refer to the same entity Ei. The proposed language captures in Fig. 4.5 the structure and semantics of the relation.

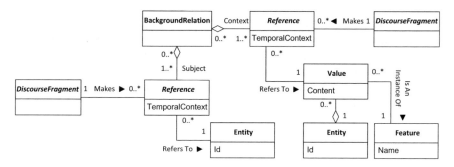

Fig. 4.5 Metamodel for the background coherence relation

Evaluation Relation

According to Hobbs' definition, an "Evaluation" relation is one of coherence between two elements of discourse in which "from S_1 infer that S_0 is a step in a plan for achieving some goal of the discourse: That is S_1 tells you why S_0 was said. The relation can also be reversed: From S_0 infer that S_1 is a step in a plan for achieving some goal of the discourse."

According to Hobbs' definition, we describe the "Evaluation" relation with the formula:

$$A \, eva \, B$$

where A and B are fragments of the discourse which play different roles in the Evaluation coherence relation. A plays the role of "content", whereas B plays the role of "motivation". Therefore, the semantics associated with this relation is that B explains why A is present in the discourse. The proposed language captures the structure and semantics of the relation as shown in Fig. 4.6.

Explanation Relation

According to Hobbs' definition, an "Explanation" relation is one of coherence between two elements of discourse which indicate a causal relation: "Infer that the state or event asserted by S_1 causes or could cause the state or event asserted by S_0."

According to Hobbs' definition, we describe the "Explanation" relation with the formula:

$$A \, exp \, B$$

where A and B are fragments of the discourse which play different roles in the Explanation coherence relation. A plays the role of "cause", whereas B plays the role of "effect". Therefore, the semantics associated with this relation is that A causes B and that this causality may be implicitly or explicitly contemplated in the discourse. The proposed language captures the structure and semantics of the relation as shown in Fig. 4.7.

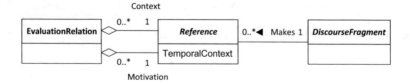

Fig. 4.6 Metamodel for the evaluation coherence relation

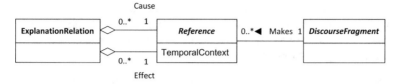

Fig. 4.7 Metamodel for the explanation coherence relation

Parallelism Relation

According to Hobbs' definition, a "Parallel" relation is one of coherence between two elements of the discourse with a structural relation between them: "Infer p (a1, a2, ...) from the assertion of S_0 and p (b1, b2, ...) from the assertion of S_1, where ai and bi are similar, for all i. Two entities are similar if they share some (reasonably specific) property."

According to Hobbs' definition, we describe the "Parallel" relation renamed it as Parallelism relation with the formula:

$$\text{par Ei Vij}$$

where Ei are fragments of the discourse which refer to entities Ei of the same type. Vij are fragments of the discourse which describe values of a set of properties of Ei. Therefore, par Ei Vij corresponds to a table in which each row corresponds to an entity Ei, each column represents a property Fj, and each cell contains a value Vij of this property for each entity Ei. Semantically, the "Parallelism" relation establishes a structural organization of the discourse which allows us to compare the values of the properties in different entities, thus permitting new inferences to be developed. These inferences may be found explicitly in the discourse or not. The proposed language captures the structure and semantics of the relation as shown in Fig. 4.8.

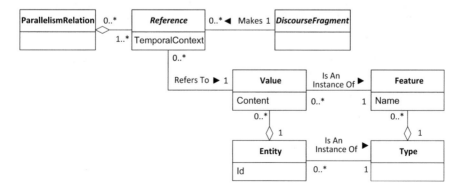

Fig. 4.8 Metamodel for the parallelism coherence relation

However, the metamodel of the "Parallelism" relation shown in the figure above is not enough to demonstrate the semantic implications of this relation in the discourse. For this reason, two OCL expressions have been added in order to increase the degree of formalization of the metamodel:

- Firstly, the entities which participate in a parallel relation should belong to the same type:

```
context ParallelismRelation
self.Reference->forAll(
  r1, r2, r1.Value.Entity.Type =
  r2.Value.Entity.Type)
```

- Secondly, the whole of the characteristics for each entity should be the same for all the entities involved:

```
context Entity
foal(e1, e2, e1 <> e2,
  e1.Value as Set1, e2.Value as Set2)
Set1{n}.Feature = Set2{n}.Feature
```

Elaboration Relation

According to Hobbs' definition, an "Elaboration" relation is a particular case of the "Parallelism" relation explained above, in which "the similar entities ai and bi are in fact identical, for all i. It can be given the following definition: Infer the same proposition P from the assertions of S0 and S1."

```
                    elab Ei Vij
```

Therefore, the metamodel of the "Elaboration" relation is the same as that of the "Parallelism" relation but in this case we must use OCL expressions to show the restriction that the V_{ij} values for a specific characteristic are identical for all the entities E_i. To sum up, the OCL restrictions at work in the "Elaboration" relation are:

- Firstly, that the entities which participate in an Elaboration relation should belong to the same type:

```
context ElaborationRelation
self.Reference->forAll(
  r1, r2, r1.Value.Entity.Type =
  r2.Value.Entity.Type)
```

- Secondly, that the whole of the characteristics for each entity should be the same for all of the entities involved (just as in the "Parallelism" relation) but in this case we must guarantee with an OCL directive that the number of entity values is the same for all the entities involved:

```
context Entity
forAll(e1, e2, e1 <> e2,
  e1.Value as Set1, e2.Value as Set2)
Set1{n}.Feature = Set2{n}.Feature
```

```
context Entity
forAll(e1, e2, e1 <> e2,
  set->count(e1.Value) =
  set->count(e2.Value))
```

- Finally, for each characteristic involved, the content of its values should be identical for all of the entities:

```
context Feature
self.Value->forAll(v1, v2 | v1 <> v2 ->
  v1.Content = v2.Content)
```

Exemplification Relation

According to Hobbs' definition, an "Exemplification" relation is one of coherence between two elements of the discourse in which S_1 is the current discourse clause and S_0 a previous clause in the discourse. Thus, an exemplification relation is defined as: "Infer p(A) from the assertion of S_0 and P(A) from the assertion of S_1, where *a* is a member or subset of A."

According to Hobbs' definition, we describe the "Exemplification" relation with the formula:

$$\text{Vij exe Wk1}$$

where each V_{ij} is a fragment of the discourse which makes reference to a value V of a characteristic F_j of an entity E_i. Each W_{kl} is a fragment of the discourse which makes reference to a value W of a characteristic G_l of an entity D_k. It should be taken into account that G_l and F_j may make reference to the same characteristics of the same entities, coincide partially or differ completely. Likewise, the ensembles E_i and D_k may make reference to the same entities, coincide partially or be completely different.

V_{ij} are called "bases" and W_{kl} "examples". This means that one or several fragments of the discourse, playing the role of bases, may be exemplified by one or several fragments of the discourse playing the role of examples. The exemplification mechanism implies making references to values belonging to the same or different entities of the field. The exemplification is only produced when the values involved belong to the same entity, or when they belong to different entities which are strongly related. Due to the fact that exemplifying necessarily implies decreasing in the level of abstraction (that is to say, going from an abstract concept to a more specific one), the relations which make this possible should be those which implement an abstraction/precision connection. As can be seen in [31], these relations which decrease the level of abstraction have been characterized as "classification/instantiation", "generalization/specialization" and "whole/part". These relations between the entities in an exemplification relation are shown using the "ExemplificationEntityRelation" class. This class encapsulates what type of relation between entities is shown in a specific case of exemplification. In Fig. 4.9, the proposed language shows the structure and semantics of the relation.

Generalization Relation

According to Hobbs' definition, a "Generalization" relation is one of coherence between two elements of the discourse in which S_0 and S_1 are inverted compared with the "Exemplification" relation.

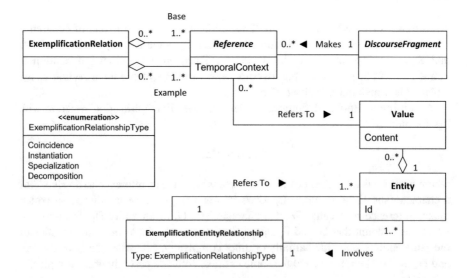

Fig. 4.9 Metamodel for the exemplification coherence relation

According to Hobbs' definition, we describe the "Generalization" relation with the formula:

$$Vij \ gen \ Wk1$$

where each V_{ij} is a fragment of the discourse which makes reference to a value V of a characteristic F_j of an entity E_i. Each W_{kl} is a fragment of the discourse which makes reference to a value W of a characteristic G_l of an entity D_k. It should be taken into account that G_l and F_j may make reference to the same characteristics of the same entities, coincide partially or differ completely. Likewise, the ensembles E_i and D_k may make reference to the same entities, coincide partially or be completely different.

V_{ij} are called "premises" and W_{kl} "conclusions". This means that one or several fragments of the discourse, playing the role of premises, can support one or several ideas expressed in other fragments of the discourse, playing the role of conclusions. The mechanism of generalization implies making references to values belonging to the same or different entities of the field. The generalization is only produced when the values involved belong to the same entity or when they belong to different entities which are strongly connected. Due to the fact that generalization necessarily implies an increase in the level of abstraction (that is to say, going from a specific concept to a more abstract one), the relations which make this possible should be those which implement an abstraction/precision connection. As can be seen in [118], these relations of increasing the level of abstraction have been characterized as "classification/instantiation", "generalization/specialization" and "whole/part". These generalization relations are captured using the "GeneralizationEntityRelation" class. This class encapsulates what type of relation between entities is present in a specific case of generalization. In Fig. 4.10, the proposed language shows the structure and semantics of the relation.

Contrast Relation

According to Hobbs' definition, a "Contrast" relation is one of coherence between two elements of the discourse which (1) present predicates of contrast made on similar entities or (2) present the same predicate regarding entities which contrast conceptually.

According to Hobbs' definition, we describe the "Contrast" relation of type 1 with the formula:

$$VAj \ con1 \ VBj$$

In this first case of the "Contrast" relation, VA_j and VB_j are fragments of discourse which refer to different values of the same characteristics of two entities. The entities have a relation between themselves which is expressed by an increase or a decrease in the level of abstraction. As has previously been seen, the relations

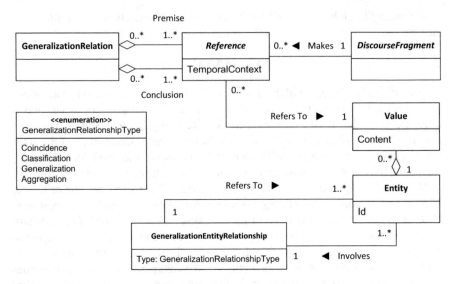

Fig. 4.10 Metamodel for the generalization coherence relation

which make this possible should be those which implement an abstraction/precision connection. As can be explained before [118], these relations have been characterized as "classification/instantiation", "generalization/specialization" and "whole/part". These relations are captured using the "ContrastEntityRelation" class. This class encapsulates what type of relation between entities is present in a specific case of contrast. In Fig. 4.11, the proposed language shows the structure and semantics of the relation.

In addition to the fact that the entities are similar, we must use OCL expressions to capture the fact that the values V to which both fragments of the discourse refer are different. The similarity of the entities and the difference between the values of the elements are what establish the semantic contrast. Thereby, the OCL restrictions, which are at work in the type 1 "Contrast" relation, are:

- Firstly, the whole of the characteristics for each entity has to be the same for all the entities involved:

```
context Entity
forAll(e1, e2, e1 <> e2,
  e1.values as Set1, e2.values as Set2)
Set1{n}.feature = Set2{n}.feature
```

- Secondly, for each characteristic involved, we have different contents in the value of that characteristic:

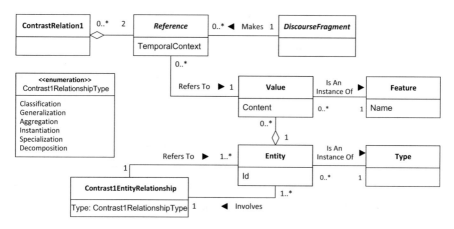

Fig. 4.11 Metamodel for the type 1 contrast coherence relation

```
context Feature
self.value->forAll(v1,v2 | v1 <> v2 ->
  v1.content <>v2.content)
```

As for the type 2 contrast relation identified by Hobbs, we can describe it with the formula:

$$VA_j \ con2 \ VB_j$$

In the second case, VA_j and VB_j are fragments of discourse which refer to the same value of the same characteristics of two entities. However, the entities which take part in the contrast relation present a conceptual contrast in their very nature, in their definition or in some aspect related to the reality which they represent, thus establishing semantics of contrast in the discourse. In Fig. 4.12, the proposed language shows the structure and semantics of the relation.

It should be noted that entities A and B belong to the same type as they must show common characteristics. In addition, we must use OCL expressions to capture the fact that the values of V to which both fragments of the discourse involved in the relation refer are the same. Therefore, the OCL restrictions which are at work in the type 2 "Contrast" relation are:

- Firstly, the entities which take part in a contrast relation (case 2) should belong to the same type:

```
context ContrastRelation2
self.Reference->forAll(
  r1, r2, r1.Value.Entity.Type =
  r2.Value.Entity.Type)
```

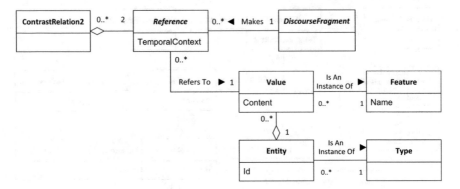

Fig. 4.12 Metamodel for the type 2 contrast coherence relation

- Secondly, the whole of the characteristics for each entity has to be the same for all the entities involved:

```
context Entity
forAll(e1, e2, e1 <> e2,
  e1.Value as Set1, e2.Value as Set2)
Set1{n}.Feature = Set2{n}.Feature
```

- Finally, for each characteristic involved, we have the same content in the value of that characteristic:

```
context Feature
self.Value->forAll(v1,v2 | v1 <> v2 ->
  v1.Content = v2.Content)
```

Violated Expectation Relation

According to Hobbs' definition, a "Violated Expectation" relation is a particular case of the "Contrast" relation, in which there is only one entity involved and the contrast situation is between the real and expected values for certain properties of this entity. In the words of Hobbs: "Infer P from the assertion of S_0 and P from the assertion of S_1."

According to Hobbs' definition, we describe the "Violated Expectation" relation with the formula:

$$Vij \; vex \; Wij$$

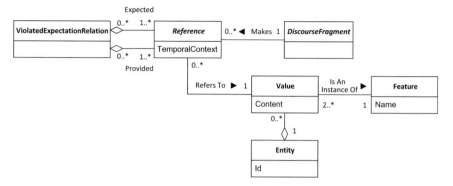

Fig. 4.13 Metamodel for the violated expectation coherence relation

where V_{ij} and W_{ij} are fragments of the discourse which refer to the values of V and W of characteristics of the same entity. This coherence relation indicates that the values of W contradict an unexpressed perception in the discourse which is generally assumed when we analyze V or when some of the characteristics of the entity are evaluated. In Fig. 4.13, the proposed language shows the structure and semantics of the relation.

It should be noted that the values V and W should belong to the same entity, thus making necessary the OCL restriction:

- The values of the entities involved should belong to the same entity:

```
context ViolatedExpectation
self.Reference->forAll(
 r1, r2, r1.Value.Entity =
 r2.Value.Entity)
```

The Methodological Aspects

According to the literature reviewed, the process of discourse analysis is generally carried out ad hoc, albeit with some exceptions explained above, such as its application in certain legal texts [187]. In no case does this process have a formal mechanism to precisely express, without ambiguity, the structure of a specific text and the semantics contained within it: The modeling language for discourse analysis described in the previous section allows us to structure and extract the semantics contained in textual sources written in freestyle, mapping the elements of both the discourse and the field in question and identifying the coherence relations defined by Hobbs [120]. It, therefore, becomes a formal mechanism for analyzing textual sources written in freestyle, independently of the field of application in which we are moving. However, if we wish to completely integrate the process of discourse analysis as part of any software engineering methodology with the need to analyze textual sources, it is necessary to establish the relation between the formal mechanism created and other methodological elements (tasks, agents, products, etc.) involved in this process. This section describes the methodological elements necessary in order to integrate discourse analysis into the process of software development, along the lines of the ISO standard *ISO/IEC 24744*, which has previously been outlined.

Hobbs describes the process of discourse analysis as an iterative process in which, in order to analyze any textual source, four steps should be taken:

In the first place, long fragments of text (typically paragraphs) are identified which maintain a semantic coherence (a global idea, a theme, etc.). The identification of these large units of text, known as fragments of discourse, in an iterative process in which the groups obtained in the first iteration are analyzed recursively in order to identify fragments inside them until the smallest possible units (typically sentences or individual noun phrases) are obtained.

Secondly, each pair of *syntagmas* or sentences is labelled with a particular coherence relation. Hobbs' method includes coherence relations such as causal arguments, consequence relations, contrast arguments, exemplifications and generalizations of arguments. The choice of coherence relation for any given pair of discourse fragments is based on two aspects:

- The structure of the clauses which participate in them: in the afore-mentioned language it is possible to see how each one of the coherence relations responds to a specific structure, which we have captured in a metamodel. This structure and certain other elements, such as the presence of certain key words or grammatical connectors, determine the choice of one coherence relation or another for each pair of discourse fragments. For example, the presence of causal grammatical connectors generally indicates that the causal coherence relation is a good candidate to label that pair of discourse fragments.
- The semantic references which the clauses make to elements of the field. For example, if the two clauses describe an entity which undergoes a change in time, it is extremely probable that one of the coherence relations implying a temporal change is a good candidate to label those two discourse fragments.

The third step is to identify the components which make up the coherence relation chosen as the label within each pair of associated discourse fragments. Each particular coherence relation follows a given formal structure with a well-defined set of components. For example, it is expected that all Occasion coherence relations refer to one or more entities in two different temporal situations; the entity (or entities) and the two temporal situations must be components identified within the pair of discourse fragments labelled with the Occasion coherence relation. Using the language created during this doctoral research project, these structures have been formalized in the form of metamodels, thus making it easier to identify and to verify whether a pair of discourse fragments respond to one particular coherence relation.

Steps two and three are carried out recursively with a *bottom-up* focus, taking simple clauses as a starting point and following the structure of the breakdown of the discourse "upwards" until the longest fragments of discourse have been analyzed. Therefore, coherence relations are not only established between clauses but also between larger units of discourse, such as simple phrases, complex phrases and even between whole paragraphs.

Finally, Hobbs' approach includes a fourth validation step. A common practice in validating the results of an analysis of a discourse is to evaluate the conclusions with experts in the field in question or, if possible, with the original author of the text.

Following the specifications of the process of discourse analysis defined by Hobbs, we have defined a complete methodology expressed in the ISO standard *ISO/IEC 24744*, which formally encapsulates all the methodological elements necessary to carry out an analysis of discourse in software engineering contexts and using our own modeling language, which has been explained above.

Figure 4.14 provides an overview of the process of discourse analysis expressed in a *ISO/IEC 24744* process diagram. In this way, the process of discourse analysis is divided into four tasks, which correspond to the four steps identified by Hobbs. In addition, the diagram of the process shows other details, such as the agents involved in each task (*producers* in the terminology of *ISO/IEC 24744*. The fourth step, the task of validation, is represented by a plus sign, which indicates that it is a recommended, though not obligatory, task in order to complete the process of discourse analysis.

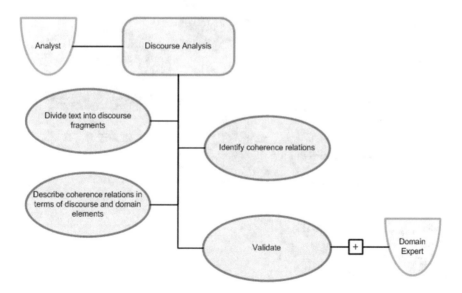

Fig. 4.14 ISO/IEC 24744 process diagram representing the process of discourse analysis

Furthermore, the ISO standard *ISO/IEC 24744* defines an action diagram to complete the full representation of the described methodology. This type of diagram also shows the tasks involved in the defined process but in relation to the products (source, intermediate or end) implied in the proposed methodology, as well as the different actions which must be carried out in order to obtain the said products (a product can be read, created, modified, etc.).

Figure 4.15 shows the action diagram defined to methodologically represent the process of discourse analysis with two products identified within it: the specific textual source to be analyzed and which is read in different tasks and the discourse model created by using the proposed modeling language. This discourse model is created in the first task and continuously evolves throughout the whole process of the discourse analysis. The discourse model as an end product allows us to add structure and maintain the underlying semantics in the text of origin.

It should be noted here that the use of the ISO standard *ISO/IEC 24744* to describe the methodological components involved in the analysis of discourse allows for the inclusion of this methodology in any software engineering methodology which can be expressed following the afore-mentioned standard. Therefore, the methodology presented here can be inserted into those parts of the software development process in which it is necessary to structure and analyze the semantics of texts written in freestyle. In the same way, the choice of this standard for the origin

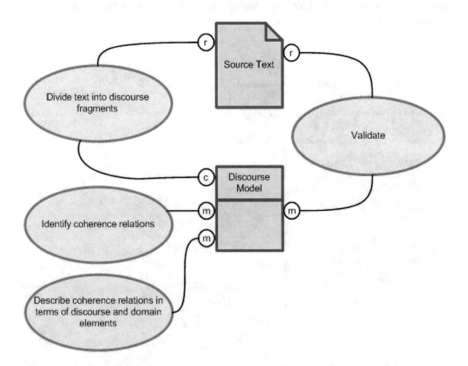

Fig. 4.15 The ISO/IEC 24744 action diagram representing the methodology of discourse analysis

of the language and the definition of the methodology allows the whole solution to be independent of the field of application, enabling it to be used to analyze texts from other fields and to integrate the process of discourse analysis in other specified processes which follow *ISO/IEC 24744*. The application of the methodology described here, including the modeling language, for archaeological discourse analysis allows us to provide structure and to extract both semantic and inferential information from textual sources in the field, thus constituting our method of information extraction regarding how knowledge is generated in archaeology. This application shall be described in later sections contained in Part III.

Archaeological Cognitive Processes Particularities

A Characterization of Cognitive Processes in Archaeology

The Proposed Characterization

As can be seen previously (see Chap. 3 for the entire review), the existing characterizations work on very different levels of abstraction, have arisen from within the context of various disciplines and have been applied to very different contexts. The choice of one characterization or another to be used in the software assistance to knowledge generation in archaeology will, therefore, be determined by the following principles:

- Although the software assistance to knowledge generation which we are seeking lies within archaeology, we believe that the chosen characterization should work on a medium-high level of abstraction, which will not allow us to refer to extremely specific cognitive tasks and should be as independent from the field of application as possible.
- The chosen characterization should, however, adapt itself to the particularities which have been detected previously in this chapter as far as the generation of knowledge in archaeology is concerned, especially in terms of data [97] and the predominance of textual sources [171].
- Although the software assistance to knowledge generation will show itself in terms of adapted visualizations, the solution that we propose should allow for the later incorporation of other assistance methods, especially knowledge extraction techniques. For this reason, the chosen characterization should be defined not only in terms of visualization primitives, but also of true primitives of the cognitive processes to be assisted.

Having analyzed these principles, we believe that it is appropriate to take the characterization described by Hobbs as a basis, given the fact that it fulfils the three criteria highlighted above (a certain degree of independence from the discipline of archaeology and an average degree of abstraction, adaptation to the characteristics

of the textual sources being handled and vague in terms of the definition of the
existing primitives of information visualization). What is more, we believe that the
fact that this characterization of cognitive processes has previously been used in
software contexts in the fields of biomedicine and legal texts proves its flexibility in
terms of fields of application, having a strong presence in textual sources and its
appropriateness for use in software contexts. Having said this, it should be pointed
out that, as far as we have been able to discover during the writing of this book,
Hobbs' characterization has not been included in any framework whose objective
has been to provide software assistance, but has been used in more automatic
contexts [153, 187].

Therefore, the coherence relations defined by Hobbs serve as a basis for the
characterization of the cognitive processes in which we aim to assist. However,
following several iterations with real users, which will be described in Chap. 8, the
coherence relations were organized into four groups, due to the fact that the need
arose to include semantics related not only with the specific cognitive process being
assisted but also related with the aim of the archaeologists when carrying out this
process. The advantages regarding the flexibility of the definitions of cognitive
processes on various hierarchical levels have been dealt with by several of the
authors mentioned above [170].

In the end, the final characterization was organized into two hierarchical levels,
one corresponding to the coherence relations defined by Hobbs and another,
designed for the purposes of this research, grouping the coherence relations around
the final objective of the archaeologist. The definitions of the primitives of the level
created are:

- Building: inferences based on the structure of the data, rather than on particular
  values.
- Clustering: inferences based on the grouping of the data.
- Situating: inferences based on situating the data in a particular context or setting.
- Combining: inferences on the basis of the combination and/or the comparison of
  the values of the different elements of the data.

The complete characterization can be seen in Table 4.1.

This proposed characterization shall be used as a basis for the formalization of
the cognitive processes carried out by archaeologists, in such a way that they
constitute the cognitive processes involved in the process of knowledge generation
which we aim to assist in this area via the use of software.

It should be noted that the proposed characterization constituting the scientific
contribution of this research, in spite of the fact that it has arisen as a result of
analyzing the specific field of archaeology, has been expressed as primitives
independently of the field of application. This decision can be attributed to two
fundamental reasons (1) the field of Cultural Heritage draws together different
disciplines which require their cognitive processes to be characterized independently of the sub-discipline involved and (2) we believe that, although our contribution will only be validated via this research in archaeology, this approach

Table 4.1 Proposal for the characterization of the most common cognitive processes in the generation of knowledge in archaeology

Coherence relations (Hobbs)	Type of inference according to the objective of the specialist
Parallel (termed Parallelism in this book) Elaboration Exemplification Generalization	Building
Contrast Violated expectation	Clustering
Background	Situating
Occasion Explanation Evaluation	Combining

(independent of the domain) could serve to take these primitives as a basis, and validate and/or adopt them as a reference point in later studies involving software assistance to cognitive processes independently of the field of application or the case studies presented. Now, we shall go back to the initial objective and explain in detail the application of the proposed characterization to archaeology.

The Application of the Characterization Obtained to Archaeology

The characterization presented above serves as a basis for the formalization of the cognitive processes carried out by archaeologists. But how can we apply this characterization to software assistance in the domain?

We can take Chen's model outlined above as a reference for software assistance to knowledge generation. As was explained above, Chen's pipeline model for software assistance to knowledge generation is a framework which captures information on the cognitive process being carried out by the specialist in the field in question and uses it to adapt the way the system works. In this point, we propose to integrate the characterization of cognitive processes, as described above, into this model. The complete process can be seen in Fig. 4.16.

What would the complete process of software assistance in carrying out these cognitive processes be like? Let us consider an example:

Let us imagine that a large data set regarding an archaeological site is available to a group of archaeologists. These specialists can interact with this data in many ways: by comparing different attributes of the data, by grouping data, establishing causal relations, searching for bibliography relating to this data, etc. All of these possible behavioral patterns of the users are captured in the system and are described in terms of the different categories of cognitive processes and coherence

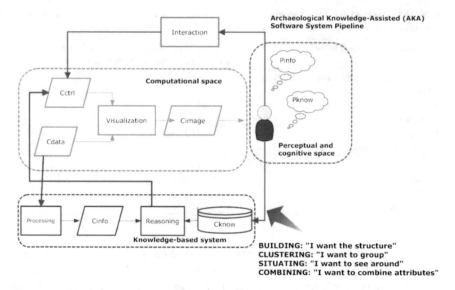

Fig. 4.16 The proposed model for assisting to knowledge generation, based on Chen's model, including the characterization of cognitive processes in archaeology proposed here

relations. Thus, during an interactive session with a user, this model is used so that the system can adapt as needed. For example, if a user is attempting to verify the correlation between two attributes, the system will identify this task as belonging to the Combining category and possibly to the Evaluation coherence relation, according to Hobbs. So, the system adapts to the user by presenting the most appropriate elements of the user's interface and the visualization options for this particular type of task.

The assistance lent by this pipeline model allows archaeologists and related professionals to carry out analytical tasks on structured data and to be more aware of why and how they have reached this particular piece of knowledge: What data has been used as a basis in the generation of this knowledge and/or what types of reasoning have they used? This self-knowledge of their discipline's cognitive processes and their application is not possible through the use of traditional methods.

In addition to studies which have the objective of characterizing cognitive processes, there are archaeological works which show the possible ways in which this kind of characterization of cognitive processes can be applied, not only referring to software-assisted knowledge generation (the fundamental goal of this volume) but also in other related areas, such as those focused on the automation and simulation of social processes from the past (Barceló, etc.). The characterization presented in this section opens new possibilities for this type of approach.

How to Validate Empirically Cognitive Processes in Software Assistance Contexts

Although the characterization of cognitive processes detailed in the previous section arose within the near contexts to archaeology and their extraction process included empirical activities, their validity in the field must be confirmed. This need for validation gave rise to a bibliographic review of the methods used to validate this type of characterizations in software engineering (in areas with cognitive necessities such as the usability and extraction of software requirements) and in other disciplines which provide methods for the validation of such characterizations, such as sociological or psychological studies.

In this section, we shall describe reference works and the reasons why the decision was made to use Thinking Aloud Protocols [162] (hereinafter TAP) in order to validate the existing characterization. Later, we shall go on to describe the design of the validation model created based on TAP, with the aim of validating cognitive processes in software assistance contexts. This design represents a scientific contribution in itself, given that, at the time the review of the studies listed here was carried out, no bibliographic sources were found documenting empirical TAP studies with this aim, although some were found in other contexts of software engineering. The proposed design was applied in order to validate the characterization of cognitive processes proposed for archaeology, the empirical studies of which are described in the following chapter.

The Validation of Cognitive Processes in Information Systems

There are many methods of very different types for performing validations in software engineering, ranging from scenarios or case studies [219], to statistical experimentation [284], prototyping and concept tests [249]. Within the types of validations with an empirical component, approaches based on observation, documentation and formalization of the resolution of problems or tasks are the most common [152, 284].

Due to the nature of the product to be validated (the characterization of cognitive processes in software assistance contexts), it was necessary to select techniques which would allow us to obtain as much information as possible regarding how the cognitive processes were carried out during this validation work. This necessity led us to consider the TAP protocol as a possible basis for the validation of the proposed characterization.

The term TAP (Thinking Aloud Protocols) is applied to a set of techniques which originated in the field of experimental cognitive psychology, in which experts in the field being studied are asked to voice the words in their minds or, in other words, to manifest each idea that they think of when carrying out certain selected tasks which they must perform. Although this technique had been used in the past, it was in the 1980s that its use became generalized, especially in connection to studies such as that of Ericsson and Simon [76], who developed the use of these techniques for studying high level cognitive processes, especially those relating to memory, from a scientific and methodologically rigorous perspective. Years later, studies such as those by Olson [200], indicated that the use of TAP is one of the most effective ways to evaluate the highest level of cognitive processes (for example, those involving working memory) and that the technique could also be used to study the cognitive differences between individuals when carrying out the same task. Ericsson and Simon [76] concluded that, although the vision of cognitive processes provided by a TAP session may prove incomplete, the results represent a reliable source as far as the cognitive processes being carried out are concerned. Due to these ideas, TAP protocols have become common as a technique employed in research contexts studying areas which are traditionally verbal or narrative, such as psychological or linguistic [200] and idiomatic [57, 156] studies. They are also commonly used in problem-solving contexts in different fields, such as physics and mathematics or biomedicine [244, 271]. These two approaches (more textual and/or narrative or more focused on problem-solving) employ their own application methodologies of the protocol.

The studies found which use TAP protocols in archaeological or heritage areas are worthy of special mention. The majority of them employ the protocol as the base methodology for ethnographic work or for interviews relating to tourism [48] but they do not characterize the cognitive processes which are performed during the sessions. Therefore, their application of the TAP protocol is of no use to us as a basis for our objective. Studies of humanities students regarding their temporal reasoning [86, 285] or the extraction of specific sub-processes, such as the transcription of ancient texts [259], are the only reference found regarding the application of TAP for an objective similar to our own, showing satisfactory results in both sub-disciplines.

As for how and why TAP protocols have been used in software engineering, there are numerous studies which extract cognitive processes during the carrying out of modeling tasks [216], decision-taking in programming [36] and prototyping and validation tasks [122, 157, 195, 273]. Furthermore, TAP protocols have been used for tasks more closely related to the extraction of cognitive processes during software-assisted knowledge generation, such as in recommender systems [290], on-line search tasks [274], ontological work [80] and the creation of knowledge [286]. However, there is no defined and agreed upon methodology for the application of TAP in the validation of characterizations of cognitive processes in software assistance contexts. The methodological consensus regarding the application of TAP in software engineering or its sub-disciplines is more developed in visualization contexts, as we shall see further on.

In general, "the literature of think-aloud research shows its strong theoretical foundation and confirms its value as a way of exploring individuals' thought processes" [59]. In addition, it gives satisfactory results in its application in software engineering contexts, thus confirming its value as a way of exploring cognitive processes [59]. The absence of a specific methodology for the application of

TAP protocols to validate characterizations of cognitive processes in software assistance contexts, as well as the need to resort to a hybrid context combining the experience of TAP application in textual and/or narrative contexts and in problem solving, makes it necessary to design a specific method for applying the TAP protocol in order to validate characterizations of cognitive processes in software assistance contexts. This design would allow for a method of reference, which would aid researchers in the validation of the cognitive processes they have identified as relevant in their field of application, allowing them then to integrate them in the software assistance process which they may design.

Validation Model Based on Thinking Aloud

As other authors have established [59], before designing a plan of application involving TAP protocols, researchers must decide which type and what level of tasks will be carried out during the sessions, the appropriate dynamics for the sessions, the use of other possible data to support inferences based on the results of the TAP sessions and the method of analysis of these results. Below, we shall outline the literature found on each aspect and describe our own choice with a view to validating our characterization of cognitive processes proposal in software assistance contexts.

Tasks

One of the most important aspects to be taken into account is that of selecting the tasks which the participants in the TAP sessions are to carry out. Requiring tasks with a high cognitive level could lead to interference in their verbalisation [9]. For this reason, many authors recommend the selection of tasks of a simple or intermediate level of difficulty, with a certain verbal load and organized into ascending levels of difficulty [9].

Tasks involving a software assistance system show a high level of heterogeneity as far as their number and level of difficulty are concerned. Therefore, in this point, it is considered necessary to determine at least one specific task for the validation of each cognitive process present in the characterization which is to be validated. This allows the degree of possibilities to be reduced, along with the confusion associated with them, in the defined tasks and the cognitive processes which are to be validated. In addition, the tasks defined should arise spontaneously from the direct observation of everyday work in the field being validated. That is to say, they should be tasks which the participants carry out (independently of the method and with different tools) as part of their habitual work in the discipline. In our case, they should be tasks relating to the generation of knowledge carried out by archaeologists as it is these tasks that we wish to assist via the use of software. The tasks selected should be of a low level of difficulty.

The Dynamics of the Sessions

There are a large number of approaches relating to the way data is to be extracted during TAP sessions. The majority of them advocate recording the sessions though minimizing the presence of recording equipment and locating it, along with the person conducting the sessions, beside the participant or at a certain distance, not in front of them, in order to minimize the degree of intimidation of the participant [197]. Other authors complement the sessions with questionnaires [9] or include control groups which carry out the same tasks outside of the TAP session [76].

Another important aspect regarding dynamics is the need, or lack thereof, for training the participants and/or giving them prior explanations. Ideally, the participants should not require previous training [59], although on occasions they may be offered a prior orientation session in order to reduce tension levels at the start of the session (the so-called "cold start effect") [91].

As far as the number of participants is concerned, TAP sessions generally involve a reduced number of participants, basically due to two reasons: (1) The objectives of the sessions are normally qualitative, so it is not necessary to carry out a huge number of sessions in order to obtain results and (2) The great and time-consuming workload which the design of the sessions implies.

In our case, we opted to record the sessions, attempting to minimize the degree of intimidation to the participant as we believe that they are a valuable source of information for future study in the discipline. Furthermore, some simple instructions were given to the participants in the sessions regarding the characterization of the processes being validated and the tasks to be carried out, although they were not trained in TAP techniques beforehand. The decision was also made to carry out a reduced number of sessions.

The Analysis of TAP Results and Complementary Data

Finally, it is necessary to take into account aspects related to the analysis of the results when designing TAP sessions. One of the most relevant aspects is the choice of the subject of the tasks to be performed in the sessions as this will, later, affect the results. In this context, studies [59] demonstrate the effectiveness of the choice of a specific case study to be carried out during the TAP sessions as a way of, later, being able to interpret the results from a general and applied point of view.

Furthermore, the subsequent method of analysis and enrichment of the TAP results varies notably. Charters carried out a review in which the majority of the

TAP sessions analyzed combine qualitative focuses with a certain quantitative classification of the results. Although there are also purely qualitative studies [57], the presence of quantitative indicators is a constant in the analysis of results [224], obtaining them by way of answers to interview questions. However, a certain lack of structure should be maintained in some parts of the sessions as this permits valuable data to be obtained which is impossible in an entirely structured format [85].

In our case, we opted for a quantitative and variable-based analysis of the data obtained during the TAP sessions (see following chapter), although the recording of the full session allowed us to create and gather data in less structured conditions during the session itself, which we believe may be of interest for future studies in this area.

Taking into account all of these aspects, a validation model of cognitive processes in software assistance contexts, shown in Fig. 4.17, has been designed.

The proposed method of validation of characterizations of cognitive processes in software assistance contexts was followed with the aim of applying it in the validation of our characterization of cognitive processes in archaeology. The design of the specific empirical studies carried out, as well as the results obtained, is contained in the following Chap. 5.

Archaeological Data Sets Visualization Particularities

Approach for Studies of Information Visualization Techniques in Archaeology

In any software system, independently of whether its main function is to assist to the generation of knowledge or not, the delicate line between the system itself and its form of "communicating" or interacting with its users is of crucial importance. This can be seen in the many areas and sub-disciplines within software engineering which are dedicated specifically to different aspects of this human-machine relationship: HCI, adaptive systems, information visualization, etc. All of these sub-disciplines have developed a significant corpus, whose main message is that the presentation and interaction which we conceptualize and design in our system plays an essential role in terms of fulfilling the objectives, functionality, usability and quality of our system. This delicate line becomes compounded when the main function of the system is more directly related to the "human" component. For example, an industrial control system with very little interaction with humans does not require the same degree of attention as adaptive software systems for medical prosthetics or those systems whose main function is to educate or train human beings. Among the objectives of the latter type of systems is attention (in some of its facets) to human beings: health, leisure, education, etc. We can consider, therefore, that software to assist to the generation of knowledge can be classified in

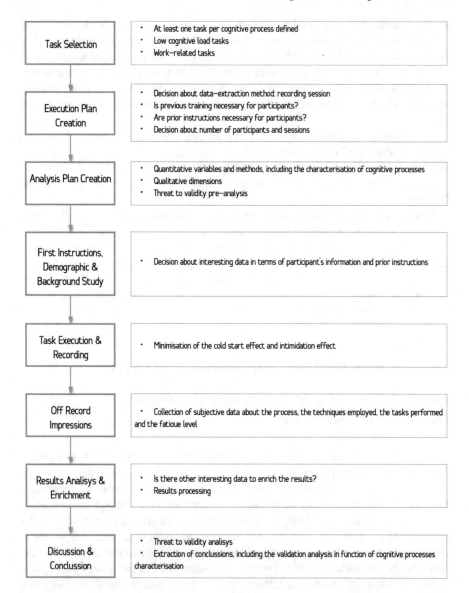

Fig. 4.17 A validation model of cognitive processes for software assistance contexts

this group as its main function consists of helping human beings to generate better knowledge based on data. In order to do this, the component of interacting and presenting information to human beings plays a fundamental role.

For this reason, we consider it necessary to empirically study which widely used and accepted visualization techniques and methods of interaction would be most appropriate for an assistance system for the generation of knowledge in archaeology. This section advances the approach taken in order to extract empirical results which may allow us to discover which methods of interaction and data presentation are most suitable for providing assistance to the generation of knowledge in archaeology.

The Proposed Studies

Two different types of empirical studies were selected. The first type consists of the application of a testbed [121] to be able to identify the existing information visualization techniques which show the best results in carrying out cognitive tasks with archaeologists, as well as allowing us to obtain information regarding the perception of them.

Testbeds have been defined as a set of artefacts associated to a software system with the necessary infrastructure for carrying out empirical studies (controlled experiments on the whole) in that system [166]. By artefact, we refer to documentation of the software, an experimental plan, test plans, the source code, different versions of the software and any other artefact which is used in the real project [166]. They are widely used in software engineering for a multitude of purposes, such as testing or other types of validations [165]. With this type of artefacts, the infrastructure of the testbed (its methods and tools) allow empirical studies to be carried out, favoring their repetition, as the studies themselves also come to form part of the testbed and, therefore, are available for use by other researchers. In our opinion, testbeds provide a straightforward, agile and flexible framework for carrying out validations aimed at obtaining quantitative results regarding one or several artefacts. For this reason, this study shall focus on obtaining quantitative information on the suitability of existing visualization techniques for large archaeological data sets.

Furthermore, as has been stressed throughout this entire chapter, TAP protocols have been widely used with good results as a tool for empirical validation in software engineering, especially as far as usability [48, 145, 157] and visualization [13, 43, 242, 246] are concerned. They have a much higher degree of methodological maturity for this objective than in the case of their use for the validation of characterizations of cognitive processes. For this reason, the second type of empirical studies will include TAP sessions which will allow us to measure the degree of suitability of certain interaction and software presentation mechanisms for the carrying out of tasks involving the defined cognitive processes. This study will also permit qualitative information of interest to be obtained via the TAP protocol.

The design of the specific empirical studies carried out, along with the results obtained from them, is described in the following Chap. 5.

Conclusions

This chapter has presented the ensemble of methods, techniques and tools which have been produced during the process of exploring the issue to be addressed, namely software assistance to the generation of knowledge, and which have allowed us to identify what needs exist and how it is possible to assist to the process of knowledge generation in archaeology via the use of software. Table 4.2 illustrates the methods, techniques and tools relevant for this issue, exploring the differences between the methods adopted (Chap. 3) and those which are original contributions (Chap. 4), along with their fields of application.

Table 4.2 Ensemble of methods, techniques and tools adopted and produced throughout this research

Method/technique/tool	Type	Description	Application area
CHARM	Adoption	Conceptual model	Cultural heritage
ConML	Adoption	Conceptual modeling language	Cultural heritage
Discourse analysis	Adoption	Technique of analysis of textual sources: structuring and extraction of semantic relations in textual sources	Independent of the field
Methodology for integrating discourse analysis in IS	Original contribution	Methodology for applying discourse analysis in software engineering	Independent of the field
Modeling language for discourse analysis	Original contribution	Modeling language for the application of discourse analysis in software engineering	Independent of the field
Characterization of cognitive processes in Archaeology	Original contribution	Identification and characterization of cognitive processes in archaeology to be assisted by software	Archaeology
Thinking aloud protocols (TAP)	Adoption	Technique for empirical experimentation	Independent of the field
Model for the validation of cognitive processes in textual sources via TAP	Original contribution	Model for validating characterizations of cognitive processes in software assistance contexts	Independent of the field
Empirical validation of visualization techniques	Adoption	Techniques for empirical experimentation	Independent of the field
Empirical studies designed for the empirical validation of visualization techniques in Archaeology	Original contribution	Testbed and TAP designed for the empirical validation of visualization techniques in Archaeology	Archaeology

The following chapter will present the results of each empirical study carried out by way of an initial validation of the methods, techniques and tools originally proposed here. Both Chaps. 4 and 5 form the basis of the definition of a solution in which software assistance to the generation of knowledge in archaeology will be dealt with in an integral way.

Chapter 5
Prior Empirical Results

Introduction

In the previous chapter, an in-depth exploration of the problem (how to use software to assist to the generation of knowledge in archaeology) was given, along with a description of existing studies in related areas. Furthermore, a narrative was provided of how a series of techniques and tools were created during the exploratory process which allowed us not only to explore, characterize and analyze the problem in more depth but also to identify what dimensions of the situation can be tackled to propose a solution. The techniques and tools created are original contributions presented here, therefore, require a prior validation in order to be able to be used as instruments in the solution to be designed. Due to this fact, this chapter deals with the empirical validation which has been carried out on each one of the original contributions, the results of which constitute empirical evidence in favor of (or against, as the case may be) the initial hypothesis, and reveals lines of action to be taken into account in order to give direction to a proposed solution for our objectives.

It should be noted that, although each of the empirical studies carried out has been designed following the usual protocols for experimentation in software engineering (the details are provided below), these studies do not attempt to establish causal relations for general application as a product of their results. Rather, they aim only to validate the suitability of the tools created for the exploration of the problem in archaeology and to confirm some intuitive lines of action with a view to designing the solution to the problem. They are exploratory solutions prior to the design of a solution and, therefore, it is not thought necessary to carry out an analysis of the results of the experimentation in terms of representativeness of the population or other similar statistical variables. Instead, the aim is to extract information on a qualitative level regarding the suitability of the techniques and tools created for our purpose and the possible lines of action. The final empirical validation which will be carried out on the proposed solution (described in Chap. 12), however, will imply

© Springer International Publishing AG 2018
P. Martin-Rodilla, *Digging into Software Knowledge Generation in Cultural Heritage*, Modeling and Optimization in Science and Technologies 11,
https://doi.org/10.1007/978-3-319-69188-6_5

empirical experimentation in which the results obtained will be statistically evaluated in an attempt to confirm the initial hypothesis and test in a real environment the solution provided.

This chapter introduces the methodology employed in the design, carrying out and analysis of the empirical studies and, later, provides the aggregate results obtained for each of the four empirical studies carried out to make an initial validation of each one of the contributions. The complete process of the design, carrying out and analysis of each study (according to the selected methodology), as well as the results of each of them, is included one by one in the detailed description of the empirical study. Thus, sometimes the chapter could be repetitive in some parts (detailing carefully the methodology and experimental reasoning after each study), but also serves the reader for reference to follow the methodology employed in a more rigorous way in their research cases. It should be noted that empirical studies have not been carried out in the case of the tools employed for the detection of conceptual needs presented in Chap. 3. This is mainly due to the fact that the chosen conceptual structure is, on the whole, an adopted contribution (ConML, CHARM). It should be highlighted, however, that both the extension and the use of certain modeling mechanisms (clusters, etc.), which shall be described in Chap. 7, do constitute an original advance of this book. In any case, it has not been considered necessary to validate these contributions empirically prior to their inclusion in a possible solution. This validation, both analytical and empirical, will be carried out when the whole proposed solution is validated. Therefore, the studies are aimed at validating the created tools for the detection of necessities regarding processes, interaction and the presentation of information detailed in previous Chap. 4.

The Design of the Empirical Studies

As other authors have pointed out [284], empirical validations in the field of software engineering can prove complex, due to the large number of aspects to be taken into account. In addition, the interdisciplinary nature of the research which concerns us here increases this complexity to a certain degree. Therefore, and in order to guarantee the rigor of all the studies and empirical validations carried out throughout the course of this doctoral research, a reference framework for experimentation in software engineering [284] has been selected which allows us to design with precision all the aspects to be taken into account in the validation and guides us throughout the entire process.

According to Wohlin, any process of experimentation in software engineering begins with an informal idea which leads us to sense that carrying out empirical studies or may be appropriate in order to verify whatever research hypothesis it is which concerns us. Taking this initial idea as a starting point, the process of experimentation in software engineering can be defined in five phases:

- Scoping: In this phase, the scope of the empirical study is marked out and the problem and general aim of the research in which the study is set is clearly defined, along with the specific objectives of the study. These specific objectives are defined by way of a template with five sections: object(s) of study, purpose, quality focus, perspective and context. For this purpose, Wohlin establishes a template in which the objective is defined in the following terms:

  Analyze the <Object(s) of study> for the purpose of <Purpose> with respect to the <Quality focus> from the point of view of the <Perspective> in the context of <Context>.

  All of these shall be explained and defined for each study in following sections. Planning: this phase proceeds to the complete design of the study to be carried out, the instruments are defined and the possible threats to the validity of the study are established. This phase has seven stages:

- Context selection, in which the context in which the empirical study will be carried out is described (on-line or off-line according to whether it is carried out in its real context of industrial and/or professional application), along with who will carry it out (students or professionals), if it is based on a *toy example* or is performed with real objects and if the study is valid for a specific or a general context.
- Hypothesis formulation, in which the null and alternative hypotheses for the planned study are formulated.
- Variables selection, in which the independent and dependent variables in the study are defined.
- Selection of subjects, in which the characteristics of the participants in the study are described.
- Choice of design type, in which the number of factors which are to be studied is chosen.
- Instrumentation, in which the objects of the study are designed.
- Validity evaluation, in which a prior analysis of the possible threats and risks of the study which has been designed is carried out.

This phase condenses the most important part of the process of experimentation as it establishes the steps to be followed in the subsequent stages, laying the foundations for the possible conclusions which can be reached by analyzing this experimentation.

Operation: this corresponds to the carrying out of the empirical study itself, in which the data is gathered according to the previously created design. It consists of three stages: preparation, execution and data validation.

Analysis and Interpretation: in this phase, the data obtained is analyzed in three stages: descriptive statistics analysis, data set reduction and hypothesis testing.

Presentation and Package: in this final phase, the way in which the results are to be presented is decided and the necessary reports are prepared. In this case, this chapter corresponds to the presentation of the results of each one of the four initial empirical studies carried out.

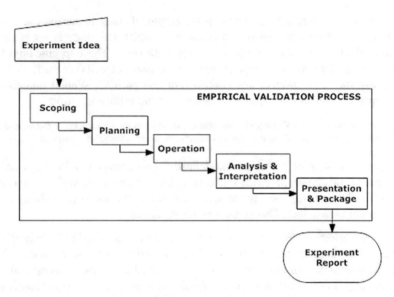

Fig. 5.1 Overview of the process of experimentation defined by Wohlin [284] and employed in this research for the design of studies and empirical validations

A summary of the process defined by Wohlin and used in this empirical validation can be seen in Fig. 5.1.

Next section details the empirical studies designed for validating the proposed strategies detailed in Chap. 4 as basic inputs for assisting archaeologists in analyzing large archaeological data sets. It is organized into sections which go into meticulous detail regarding each of the studies carried out upon testing each contribution according to this process [284].

Empirical Studies

Tools for Archaeological Process Necessities: Results

As was explained in previous chapters, the exploration of the problem on the level of process necessities led to two groups of original contributions: (1) The methodology to integrate discourse analysis into software engineering and the modeling language for the analysis of the related discourse. (2) The proposed characterization of cognitive processes in archaeology, along with a specific proposal for the validation of this characterization based on TAP protocols.

In this section, we shall examine the empirical studies which were carried out and the results obtained from them with the aim of validating each of the groups of contributions.

EMPIRICAL STUDY 1: An Analysis of Textual Sources Using our Discourse Analysis Methodology

The goal of this experiment is to obtain an empirical and intuitive idea regarding the suitability of the methodology and the proposed language associated with it for the analysis of archaeological discourse, structuring and extracting information from each fragment of discourse analyzed.

In order to do this, an empirical study was proposed in which archaeologists could execute the proposed methodology, express their degree of satisfaction with it and interact with and evaluate models created both by themselves and by other modelers regarding fragments of archaeological discourse. Thus, a corpus of 40 fragments of discourse taken from six different textual sources was selected and the proposed methodology for discourse analysis was applied, creating models using the language for discourse analysis also presented previously. Each fragment, therefore, produced its corresponding associated object model (See action model and process model diagrams in the previous chapter).

The participants had to make an initial outline of a model of the fragment of discourse, pointing out any problems deriving from the comprehension and application of the steps. Later, they were presented with a finished model corresponding to the fragment of discourse which they had attempted to model and they were asked to evaluate it and point out any problems regarding their understanding of the model or the language itself, etc. Finally, they were asked to evaluate their level of satisfaction with the methodology and with the language on a Likert scale [163] of five values. The definition and the full results of the study are described below.

Objective(s) of the Empirical Study—Scoping

This empirical study had two main objectives:

(a) On the one hand, the intention was to test the defined methodology for discourse analysis in software engineering, scrupulously following the steps designed, proving the suitability of its application to each of the 40 fragments of discourse analyzed.

(b) On the other hand, our second objective was to check whether the discourse language which was designed was appropriate for modeling different kinds of discourse (by different authors, institutions, on different topics, etc.).

According to Wohlin, these objectives can be formalized by:

(a) Analyzing the proposed methodology for the purpose of evaluation with regard to the degree of suitability to archaeological discourse analysis from the point of view of archaeologists/researchers interested in gaining a better knowledge of discourse analysis in archaeology in the context of both public and private heritage institutions.

(b) Analyzing the proposed modeling language for the purpose of evaluation with regard to the degree of suitability to archaeological discourse analysis from the point of view of archaeologists/researchers interested in gaining a better knowledge of discourse analysis in archaeology in the context of both public and private heritage institutions.

Planning the Study

Context: The empirical study was carried out with real tools and real problems. A corpus of 40 fragments of discourse was selected from six real textual sources (management and research reports in archaeology available on-line). The study was, therefore, situated in a specific but broad context with a sample of archaeologists belonging to both public and private institutions on a national scale). It could be considered that the empirical study is in an on-line context because it was carried out in the professional environment in which it would be put into practice.

The Formulation of the Hypothesis: Two hypotheses were verified. Informally, they are:

– The proposed methodology is not appropriate for discourse in archaeology. This lack of suitability should be reflected in the fact that it does not allow for agile analysis of the discourse, with many problems being identified by the participants regarding understanding of the modeling process and a low level of satisfaction among archaeologists.
– The proposed language is not appropriate for discourse in archaeology. This lack of suitability should be reflected in the fact that it does not allow for agile analysis of the discourse, with many problems being identified by the participants regarding understanding of the modeling process and a low level of satisfaction among archaeologists.

Response variables are those which formalize the variable being measured. In this case, we shall formalize the suitability of the methodology and the proposed language to discourse analysis in archaeologists by the use of two response variables for each hypothesis: the number of problems identified and the degree of satisfaction of the participant according to a Likert scale of five possible values. Based on these informal hypotheses, we can now define H_0 y H_0' as null hypotheses and their alternative hypotheses:

H_0: The proposed methodology is not suited to discourse analysis in archaeology, with numerous problems being identified by archaeologists and the degree of satisfaction being low.

The number of problems shall be considered high when, for each fragment of discourse to be evaluated, more than five problems are shown by the participant. The fragments of discourse being evaluated are small so more than five problems will generally indicate that the participant is not able to follow the methodology or use the language fluently. It should be noted that the following symbols have been used for the definition of the hypotheses:

↑ Indicates that the variable accompanying the symbol has a tendency to increase.

↓ Indicates that the variable accompanying the symbol has a tendency to decrease.

&& Indicates the union of conditions in the hypothesis.

H_0: M (Methodology) = ↑ P (Problems reported) && ↓ SD (Satisfaction Degree)
Alternative Hypothesis H_1: M (Methodology) = ↓P (Problems reported) && ↑ SD (Satisfaction Degree)
H_0': The proposed language is not suited to discourse analysis in CH, with numerous problems being identified by the specialists in the field and a ↑ low degree of satisfaction.
H_0': L (Language) = ↑ P (Problems reported) ↓ && SD (Satisfaction Degree)
Alternative Hypothesis H_1': L (Language) = ↓ P (Problems reported) && SD (Satisfaction Degree).

The selection of variables: The independent variables were selected, these were: the modeling skills and experience of the participants and those of the methodology and the language being evaluated. The variables relating to the skills of the participants would be controlled by way of a prior questionnaire. The methodology and the language being evaluated are the treatments to be applied.

The suitability to discourse analysis in archaeology, formalized by the number of problems identified and the degree of satisfaction of the participants, are the dependent variables.

The selection of subjects: The sample of participants corresponds to the *simple random sampling* model. A sample of eight archaeologists was selected, drawing them from five different public and private Spanish institutions (Incipit-CSIC, the Xunta de Galicia (the Galician regional government), private companies in the area of cultural management, the University of Santiago de Compostela and CCHS-CSIC). These participants were selected at random from the more than 20 specialists who expressed an interest in collaborating with this study.

Design Principles: As far as randomization is concerned, on an initial level, the objects were not assigned at random to the subjects. In other words, all of the participants evaluated models and their associated fragments of text from the six reports. However, on the second level, it can be considered that there was a certain degree of randomization, given the fact that the participants randomly evaluated different models from among those of each report. The objective of the study is not to evaluate this methodology and the proposed modeling language compared with other constructs. Therefore, we believe that this level of randomization is sufficient. The assignations of objects were not random as the order of evaluation was not relevant.

As far as the blocking of variables is concerned, we believe that the decision to evaluate several fragments of text from different sources and their models blocked, to a certain extent, the impact of the sources and the text itself in the resulting models. In addition, the influence of the participants' skills in modeling was taken into account, being measured by a prior questionnaire.

Finally, it would have been desirable to apply the principle of balancing but, due to the difficulty in finding a large number of subjects, they all evaluated models of fragments from six texts. Therefore, balancing did not occur.

Instrumentation: The choice of objects can be considered as random as, although some of the texts belong to the institutions for which the participants work, others form part of international institutional repositories, unconnected to the participants. In the case of reports belonging to the participants' institutions, none of the participants was the author of any of the texts.

As far as the lines of action in the execution of the study are concerned, the modeling procedure is defined by the methodology being evaluated. Instructions were given regarding the methodology and material on the language, and the participants were allowed to ask initial questions about them. It was not considered necessary to train the participants in advance about the methods to be used as the aim of the study consisted precisely in establishing the suitability of the methods in untrained conditions. Mistakes were measured and problems identified both in terms of use of the methodology and semantic problems (lack of understanding, doubts, etc.) when evaluating the model with regard to its associated fragment following the proposed language. Furthermore, questions were asked about the participants' degree of satisfaction (according to a Likert scale from Extremely Low, Low, Average, High and Extremely High) with the procedure (that is to say the methodology) and with regard to their understanding of the modeling language itself.

Evaluation of aspects of validity: According to the definition of Wohlin based on Cook and Campbell [54] with regard to types of threats in software experimentation, it was considered necessary to highlight beforehand the following threats to the validity of the study:

- With regard to its internal validity, it is considered that the low number of participants could compromise the results if the objective of the study were to establish causal conclusions. However, as was explained at the beginning of this chapter, the empirical study does not aim to establish these relations, thus minimizing this threat.
- With regard to its external validity, it is considered that the probability of the results being repeated in other contexts is high, due to the randomization of the objects used, as well as to the fact that other similar studies and experiments [172] have been carried out in heritage contexts.
- As far as the conclusion validity is concerned, it is considered that the random choice of textual reports and of the fragments to be modelled within these reports constitutes a guarantee for the representativity of the texts. Likewise, conducting sessions in person, with just one participant each time, minimized threats regarding the quality of the data obtained.
- As far as the construct validity is concerned, we should take the low level of statistical generalization presented into account, fundamentally due to the influence of threats related to social aspects, the possible variability within the evaluated models and the modeling skills of the participants. Furthermore, it would be necessary to carry out a larger number of similar empirical studies in order to avoid the problems of "*fishing and the error rate*". In other words, this study manages to identify a relation between variables which in reality does not occur on a greater scale in a higher number of studies. It should be remembered that this degree of generalization is not the final aim of the empirical study.

The Execution of the Study—Operation

Preparation: The participants did not know the fragments of text to be analyzed nor the heritage reports from which they were extracted. They were informed of the methodology and the language but not of the hypotheses being dealt with in the study. By offering themselves as volunteers for this study, they gave their consent for this. All the necessary materials were prepared beforehand: an initial questionnaire about the participants' personal and professional profile, the fragments of text selected and their associated models, information regarding the language and the questionnaire about the degree of satisfaction.

Execution: The participants were received one by one in individual sessions lasting between an hour and an hour and a half. An explanation was given to each volunteer regarding the methodology and some basic notions of the language and they were asked to sketch a model for each fragment, following the methodology and stating any doubts, problems and mistakes which they observed in its application. Later, they were shown a final model made by experts in modeling for each fragment of text. They were asked to evaluate the model, comparing it with their sketch and giving information about each semantic and/or syntactical problem which they identified. Finally, they were questioned about their level of satisfaction, both concerning the methodology and the language, as well as the model presented.

Data validation: No invalid data was detected. This can be attributed to individuality and supervision in the process of the execution of the study.

Analysis and Interpretation: Results

Results of the use of the methodology (P): The rate of problems detected in the understanding and application of the described methodology is extremely low.

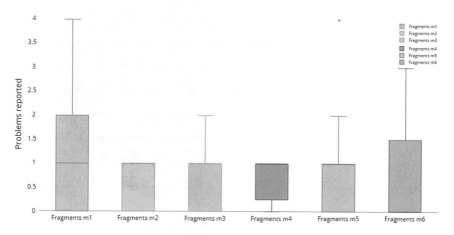

Fig. 5.2 Distribution of problems detected in the understanding and application of the methodology according to the textual fragments analyzed

As can be seen in Fig. 5.2, between 0 and 4 problems were detected in the fragments corresponding to the different textual reports, this can be put down to the simplicity of the methodology, its ease of understanding and/or application or other factors which allow us to see a level of suitability to the purpose which concerns us. It should be noted that none of the participants are specialists in software engineering (although as archaeologists could have certain skills as a basic user, it is not the case as software specialists) and, therefore, modeling methodologies are foreign to them in the course of their daily work (although some of them may have a degree of knowledge of some of their basic principles). In addition, none of the participants had prior contact with the methodology designed nor were they aware of the aim of this methodology.

As far as the **degree of satisfaction with the use of the methodology (SD)**, which the participants expressed, is concerned, more than half expressed a level of satisfaction higher than 3 (Average), as can be seen in Fig. 5.3:

Results of the discourse modeling using the proposed language (P): The rate of problems detected relating to understanding and applying the language itself increases slightly in all the reports compared to when methodological aspects are concerned. This can be put down to the fact that the language has many more elements which can cause problems of semantic understanding on the part of the

Fig. 5.3 Degree of satisfaction concerning the methodology employed, as expressed by the participants

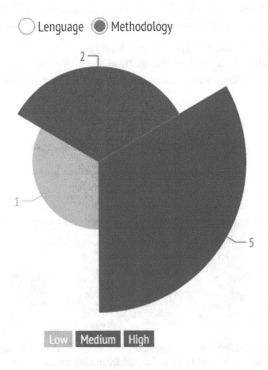

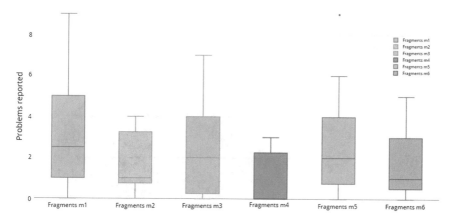

Fig. 5.4 Distribution of problems detected in the comprehension and application of the modeling language according to the textual fragments analyzed

participant, taking into account the fact that it represents a greater learning curve. Furthermore, there is a resistance to modeling following the established form of notation with participants preferring to express the casuistry of each fragment being modeling verbally during the study. The participants also asked for examples in order to understand the different definitions of the main concepts of the language. These aspects are of great relevance when it comes to proposing a solution to the problem in question (see Fig. 5.4).

As far as the **degree of satisfaction with the use of the language (SD)** is concerned, the tensions detailed above may lie behind the lower degree of satisfaction than was the case with the methodology. However, there is still a relatively high degree of satisfaction, perhaps due to the fact that each participant has had to reflect on conclusions in their own field during the study, as they themselves made clear during the sessions (see Fig. 5.5).

Presentation and Package
This section represents the presentation format for the empirical study 1.

Summary of Results and Conclusions
This empirical study enabled us to obtain an initial idea of the degree of suitability of the methodology and of the language proposed for its purpose (the modeling of discourse) and the later use of this information. However, it did not allow us to infer any generalizations or causal relations about it, which will be proved in later statistically representative studies.

It must be noted that, thanks to this study, we were able to discover some features which provide this series of techniques and tools with suitability and a higher degree of satisfaction:

- The semantics of the language should be clear and contain precise definitions and examples in the field of application (in this case archaeology) in such a way

Fig. 5.5 Degree of
satisfaction regarding the
modeling language, as
expressed by the participants

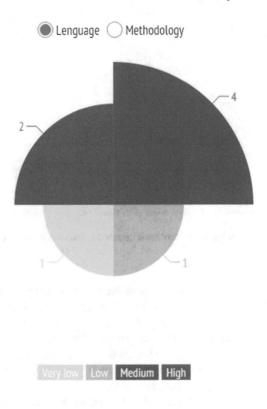

that the participants can identify when to model or positively evaluate a specific
inference present in a specific fragment of discourse.

• When observing the complete model of a fragment of discourse, the reaction is
always positive, the advantages of the structuring of the discourse and its
possibilities with a view to reflection about how knowledge is generated in the
discipline is understood.

Furthermore, the planning and execution of the empirical study enabled us to
obtain a corpus of analyzed, and later modelled, textual fragments with their
evaluation by archaeologists. This situation enabled a small analysis of the presence
of cognitive processes in the analyzed fragments to be carried out, producing an
interesting and broad sample. In Fig. 5.6 the results of this analysis can be seen.

As can be seen, all the types of cognitive processes defined in our characterization are present, with the most common being those based on combinations of
values (Occasion, Explanation, Elaboration, and Evaluation) and the changes of
levels of abstraction (Exemplification/Generalization). These relations generally
imply temporal and geographical components in the text in which they are

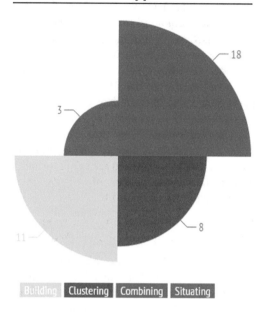

Fig. 5.6 The distribution of cognitive processes in the corpus of textual fragments analyzed

referenced. These a posteriori conclusions enable us to support the idea that our solution must deal with cognitive processes on different levels within the process of knowledge generation, in such a way that we can know which ones we should assist, and their close relation with how heritage knowledge is generated at the present time, as reflected in the analyses carried out on textual reports.

EMPIRICAL STUDY 2: The Empirical Validation of the Characterization of Cognitive Processes in Archaeology via the Use of TAP Protocols

At this point, it is not only necessary to carry out an initial validation of the methodology and the modeling language created but also to validate the characterization of cognitive processes proposed for archaeology. This is due to the fact that these cognitive processes form the foundations for the software assistance which we aim to propose as a solution. In order to do this, a study based on an original application of TAP protocols [162] was carried out.

The goal of this experiment is to evaluate whether the proposed characterization of cognitive processes was appropriate for the categorization of underlying cognitive processes in practical use in archaeology, mainly concerning the degree of agreement among specialists in the field when using this characterization. This aspect is particularly relevant when it comes to using the characterization as a basis for any solution or strategy within the scope of this book as it establishes, to a certain extent, the degree of generalization and acceptance that this characterization will obtain in the field in question. Therefore, textual sources were taken as a starting point (archaeological reports) and the underlying cognitive processes of 20 fragments of discourse within these reports were initially categorized. Then, these fragments were shown to archaeologists, along with several random possibilities of cognitive processes from which the fragments were to be categorized. At all times, they were asked to execute the TAP protocol (*thinking aloud protocols*). Thus, the participants stated out loud if they understood the task, if they understood the fragment of discourse to be analyzed and if they understood the semantics of the proposed characterization. They also expressed the reasons which led them to characterize the fragment as one type of cognitive process or another. Following all of this process, data was extracted regarding the degree of agreement reached by the specialists in the characterization of each fragment. The definition and the full results of the study are described below.

Objective(s) of the Empirical Study—Scoping

The objective of this empirical study, in Wohlin's terms, can be defined as:

(a) Analyze the characterization of cognitive process for the purpose of evaluation regarding the degree of its suitability to archaeology from the perspective of archaeologists/researchers with an interest in gaining knowledge of the functioning of this characterization in archaeology in the context of public and private heritage institutions.

Planning the Study

Context: The empirical study was carried out with real tools and problems (20 fragments of archaeological discourse extracted from management and research reports, which are available on-line). However, the characterization used can be considered as a "toy example" object as it is not commonly used by experts in this field. The study is set in a specific but broad context (a sample of archaeologists drawn from both public and private institutions on a national scale). It could be considered that the context of this empirical study is on-line as it was carried out in a professional context in which it would be used. It should be remembered, at this point, that the study was carried out using a methodology based on TAP (*Thinking Aloud Protocols*), specifically designed for the purpose of this empirical study and described in previous chapter.

The formulation of the hypothesis: a hypothesis was verified, which was informally defined as:

– The defined characterization of cognitive processes is not suitable for archaeology. This is reflected in the fact that the percentage of coincidence between the categorizations of the selected fragments of discourse among the different archaeologists is low. Therefore, the characterization employed may not be sufficiently representative, suitable, understandable or generalizable in the field.

In this case, we shall formalize the suitability of the characterization of cognitive processes for archaeology by way of one response variable: The average percentage of coincidence between the assignations of cognitive processes which the archae-ologists made to specific fragments of discourse. Taking this informal hypothesis as a basis, we can now define H_0 as a null hypothesis and H_1 as an alternative hypothesis:

H_0: The average percentage of coincidence (C_{AP}) among the assignations of cognitive processes to the selected fragments of discourse among different archaeologists in each group of fragments is less than 50%. H_0: $C_{AP} < 50\%$

Alternative hypothesis H_1: $C_{AP} > = 50\%$.

Selection of variables: The fragments to be evaluated and the associated cog-nitive processes were selected as the independent variables. In addition, experience with TAP protocols, controlled via a questionnaire prior to participation, was taken into account as an independent variable.

The dependent variable would be the suitability of the characterization of cog-nitive processes to archaeology, formalized via the average percentage of coinci-dences of assignations of cognitive processes performed by the archaeologists in order to characterize the fragments of discourse.

Selection of subjects: The sample of participants corresponds to the *Simple random sampling* model. A sample of six archaeologists was selected from three different institutions (both public and private) (Incipit CSIC, Xunta de Galicia (the regional government of Galicia) and a private company specialized in cultural management) chosen randomly from the 20 specialists who showed an interest in collaborating in this process.

Design principles: As far as randomization is concerned, the objects were not assigned randomly to the subjects. That is to say, all the participants evaluated the 20 selected fragments extracted from four different reports. The participants eval-uated the fragments of discourse and categorized them randomly. However, we believe that the order of evaluation is not relevant.

As for the need to block variables, we consider that the decision to evaluate several fragments of texts from different sources blocked, to a certain degree, the impact of the sources and the text itself in the resulting models. The influence of the participants' modeling skills was also taken into account, with an attempt to measure this aspect being made via a prior questionnaire.

Last of all, it would have been desirable to apply the balancing principle but, due to the difficulty in finding a high volume of subjects, they all evaluated the same fragments, therefore, balancing did not occur.

Instrumentation: The choice of the objects can be considered as random, due to the fact that, although some of the texts belonged to the participants' own insti-tutions, others came from international institutional repositories unrelated to their institutions. In the case of the reports from the institutions consulted, none of the participants were the authors of the texts.

As far lines of action in the execution of the study are concerned, the participants were provided with a document containing the instructions for participation, along with a questionnaire and information regarding the characterization of cognitive processes which should be used for their answers. No prior training was necessary for the participants to reply to the questionnaire.

Evaluation of aspects of validity: According to the Wohlin's definition, based on Cook and Campbell [54], it was considered necessary to highlight the following threats to the validity of the study beforehand:

– As far as its internal validity is concerned, we consider that the low number of participants in the study could compromise the results if the objective of the study were to establish causal conclusions. However, as was explained at the beginning of the chapter, the empirical study does not have the objective of establishing this type of relations, thus minimizing this threat.
– As far as its external validity is concerned, we consider that the probability of the results being repeated in other environments is high, due to the randomization in the objects used and the fact that other similar studies and experiments have been carried out in heritage contexts [172].
– As far as the conclusion validity is concerned, we consider that the random selection of textual reports and of the fragments to be characterized within these reports constitutes a guarantee of representativity of the texts. Furthermore, conducting the sessions in person, with only one participant at a time, minimized the threats regarding the quality of the data obtained.
– As far as its construct validity is concerned, we must take into account the fact that the low number of participants did not allow us to make a statistical generalization of the results. Another possible threat could be the suitability of the selected measures as the average percentage may not permit a more in-depth analysis of the data obtained. However, carrying out the studies according to the TAP protocol allows us to obtain more data in the future.

The Execution of the Study—Operation

Preparation: The participants in the study did not know the fragments of text which were to be characterized. They were given information about the characterization of the cognitive processes to be used but not about the hypotheses being tested in the study. By offering to participate as volunteers in this study, they gave their consent to these conditions. The necessary materials were prepared beforehand: an initial questionnaire about their professional profile and previous experience in TAP (None, Low, Average, and Expert), the selected fragments of text and the characterization to be used.

Execution: The participants were received one by one in individual sessions of between 45 min and one hour in length, thus avoiding problems relating to fatigue in TAP protocols [162]. Each of them was given an explanation of the characterization of cognitive processes to be used and they were told that they would be recorded according to the TAP protocol described in Fig. 19 (see Chap. 4). Then, they were asked to characterize the fragments of discourse, describing aloud the reasons for their choices and/or doubts.

Validation of the data: No invalid data was detected. This can be attributed to the individuality and supervision of the process of execution of the study.

Analysis and Interpretation: Results

Results according to percentage of coincidence (C_{AP}): Figures 5.7 and 5.8 show the participants' answers regarding the characterization of the 20 fragments of archaeological discourse. Figure 5.7 shows the fragments in which the variability of the answers is only of two different values. In other words, the archaeologists only differed in two possible cognitive processes when characterizing the fragments of discourse. Figure 5.8 shows those fragments with a greater degree of variability in the answers. In both cases, the most chosen option also corresponds to the primary

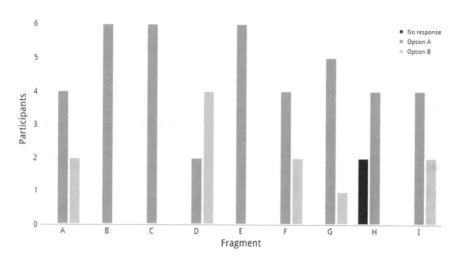

Fig. 5.7 The participants' answers for fragments A-I. As can be seen, in the majority of cases, there is a consensus as far as the cognitive process assigned to the fragment is concerned

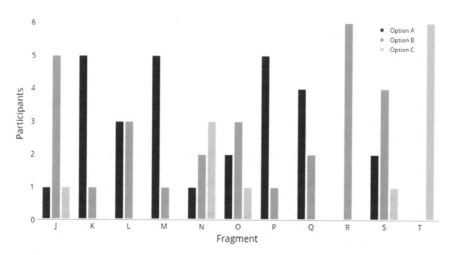

Fig. 5.8 The participants' answers for fragments J-T. As can be seen, in the majority of cases, there is a consensus as far as the cognitive process assigned to the fragment is concerned

Table 5.1 Coincidence Percentage C_{AP} obtained for each of the fragments A-T

Fragment	A	B	C	D	E
C_{AP}	66.6%	100%	100%	66.6%	100%
Main cognitive process	Explanation	Evaluation	Elaboration	Parallel	Generalisation
Cognitive process (pre-choice)	Explanation	Evaluation	Elaboration	Parallel	Generalisation
Fragment	F	G	H	I	J
C_{AP}	66.6%	83.3%	66.6%	66.6%	50%
Main cognitive process	Generalisation	Violated	Contrast	Explanation	Parallel
Cognitive process (pre-choice)	Generalisation	Violated	Contrast	Explanation	Parallel
Fragment	K	L	M	N	O
C_{AP}	83.3%	50%	83.3%	50%	50%
Main cognitive process	Evaluation	Explanation	Evaluation	Parallel	Evaluation
Cognitive process (pre-choice)	Evaluation	Explanation	Evaluation	Parallel	Evaluation
Fragment	P	Q	R	S	T
C_{AP}	83.3%	66.6%	100%	66.6%	100%
Main cognitive process	Occassion	Contrast	Contrast	Exemplification	Background
Cognitive process (pre-choice)	Occassion	Contrast	Contrast	Exemplification	Background

option taken as a reference for the author of this book, although this finding would be analyzed after the study.

As can be seen in both figures, the majority of the archaeologists coincided in the cognitive process which they associated with the fragment in question.

Table 5.1 shows the values in each step for the dependent variable C_{AP}:

It can be concluded that, although a much higher volume of participants would be needed to obtain an acceptable degree of statistical generalization, the coincidence percentage C_{AP} in the characterization of cognitive processes among specialists in the area is acceptable, passing the 50% (0.5) established as a hypothesis. Furthermore, this coincidence in all cases of the most selected cognitive process, by the author and by the specialists in the field, enables a line of action to be drawn.

Thus, the proposed characterization is established as suitable for use as a basis for the solution that we aim to propose in this book.

Presentation and Package
This section represents the presentation format for the empirical study 2.

Summary of the Results and Conclusions
This empirical study enabled us to gain an initial idea of the degree of agreement among the archaeologists regarding the categorization of certain fragments of discourse depending on the underlying cognitive process. The results of the *Thinking Aloud* empirical study showed a high degree of agreement in the community, with average percentages of coincidence of around 66%, which increases if we only deal with the first level of hierarchy of our characterization (more abstract and with only four possible values (*Situating, Building, Clustering* and *Combining*) as the greater disagreements were in the choice of sub-levels within the same level.

Thanks to the use of the Thinking Aloud Protocol, we were able to detect other aspects of interest, which, although they have not been formalized in the study, provide important qualitative information with a view to using the evaluated characterization. For example, disagreements may also occur due to the fact that the participants state that cognitive processes reflected in the text are not solidly supported or are expressed with confusion, although the characterization still allows the underlying cognitive process to be identified.

Furthermore, this study enabled us to compare the categorization carried out by experts in the field with that which was previously carried out by this book's author, whose area of expertise differs quite significantly from that of the participants in the test. In this case, the degree of agreement remained stable, maintaining a greater degree of agreement on the first level of abstraction of the proposed characterization.

Due to this fact, we believe that the characterization can work as a key element in the definition of what cognitive processes we want to provide assistance for in archaeology.

Both empirical studies 1 and 2 reaffirm our intuitive idea of the need for the treatment of cognitive aspects on a formal level in software assistance in archaeology. Due to the fact that they are hardly dealt with in the solutions we analyzed (in the Chap. 4), this formal treatment constitutes one of our main aims for improving software assistance in this field and will, therefore, form part of our proposed solution.

Tools for Archaeological Interaction and Presentation Necessities: Results

As was explained in previous chapters, the exploration of the problem on the level of interaction and presentation led to the creation of a group of original contributions: our own studies and results regarding the suitability of software assistance to knowledge generation in archaeology of certain information visualization techniques. That is to say, in addition to validating how to identify, characterize and model the cognitive processes to be taken into account in software assistance in archaeology, we should extract certain lines of action regarding what techniques of presentation and interaction are the most appropriate in order to provide assistance to these cognitive processes. Although some approaches were identified in the review which was carried out of the literature, it was considered necessary to extract empirical data directly from archaeologists.

EMPIRICAL STUDY 3: An Study of Information Visualization Techniques in Archaeology

The empirical study carried out attempted to evaluate aggregate information visualization techniques, as standard data plotting techniques, commonly used in scientific contexts but which are used infrequently in humanities contexts, especially in archaeology [172]. This evaluation has the aim of determining whether some are more suitable than others for supporting certain cognitive processes of our characterization. In addition, an attempt was made to find lines of action regarding the method and degree of interaction desired by archaeologists with these visualizations.

The choice of these types of visualizations (bar charts, bubble-based visualizations, etc.) can be attributed to the fact that preliminary studies with other, more complex and recent, information visualization techniques produced worse results [172]. These studies concluded that perhaps a lack of visual training on the part of these specialists in certain visualization strategies, along with the particularities of archaeological data, could be the reasons why these proposals presented problems. Of course, these investigations are preliminary and more work is needed along these lines in order to be able to conclude that techniques outside of those evaluated in this study show better or worse behaviors than those used. However, the search for a compromise between the specialists' familiarity with the most widely used visualization techniques and the use of innovative visualization techniques, as well as the large number of studies which relate these aggregate visualization techniques with cognitive processes referred in previous sections, led to us selecting them to verify their behavior when supporting our characterization of cognitive processes.

In order to do this, interviews were carried out with archaeologists (of different affiliations, personal and professional profiles and degrees of training in the visualization techniques in question) who were asked to carry out common data analysis tasks using these visualizations, measuring the errors and correct answers shown. In addition, they were questioned about their visualization preferences with regard to the tasks carried out. The definition and the full results of the study are described below.

Objective(s) of the Empirical Study—Scoping

In Wohlin's terms, the objective of the empirical study can be expressed as:

(a) Analyze 8 types of aggregate information visualizations with the purpose of evaluation regarding the degree of precision obtained according to the cognitive process in archaeology which they assist from the perspective of archaeologists/researchers interested in gaining knowledge about which visualizations demonstrate greater precision in this field in the context of public and private heritage institutions.

Planning the Study

Context: The empirical study was carried out partly with real tools (real archaeological data sets) but with "toy examples" (all the visualizations created are "toy examples" for this study in particular). The empirical study is set in a specific but broad context (a sample of specialists in archaeology belonging to public and private institutions on a national level). It can be considered that the study's context is on-line as it was carried out in the professional context in which it would be used.

The formulation of the hypothesis: only one hypothesis was tested. It can be informally expressed as:

– The visualization technique employed, along with its degree of suitability for certain cognitive processes in archaeology, does not have an influence on the degree of precision obtained by the participants when carrying out certain tasks. This is reflected in the mistake rate in tasks related with those processes. In other words, the mistake rate will be the same in tasks carried out with unspecific visualizations as with visualizations adapted to the cognitive process being assisted.

Based on this informal hypothesis, we can now define H_0 as a null hypothesis and H_1 as the alternative hypothesis:

H_0: with n being the visualization adapted to the cognitive process required and m being the visualizations which have not been adapted, the mistake rate MR (V_x) in tasks related to that process is the same when using m visualizations as when using n visualizations. H_0: MR (V_m) = MR (V_n)

Alternative hypothesis H_1: with n being the visualization adapted to the cognitive process required and m being the visualizations which have not been adapted, the mistake rate MR (V_x) in tasks related to that process is different when using m visualizations to when using n visualizations. H_1: MR (V_m) $\neq$ MR (V_n).

Selection of variables: The visualizations to be evaluated and the tasks to be carried out were selected as the independent variables. In addition, the archeologists' skill in the selected visualization techniques was considered as an independent variable. This was controlled via an initial questionnaire.

The dependent variable was the precision with which the tasks were carried out, formalized via the mistake rate calculated per task for each visualization obtained by the specialist. In each case, these visualizations were identified as belonging to group n (visualization adapted to the cognitive process required) or to group m (visualization not adapted to the cognitive process required).

In spite of the fact that they are not included in the study, the behavior of two other dependent variables was considered as an interesting aspect for evaluation:

- The distribution of the preference (DP) shown by the users for the visualization technique according to the cognitive process (from the first level of our characterization) which they were asked to perform.
- The general mistake rate per task carried out (MRT). This refers to the mistake rate of the archaeologist in carrying out a specific task, without taking into account the visualization employed to do this. This can enable us to find out which tasks prove more difficult for the archaeologist, independently of the visualization technique employed to carry them out.

Selection of subjects: The sample of participants corresponds to the *Quota sampling* model (This type of sampling is used to get subjects from various elements of a population). In total, the study had 16 participants (not selected randomly), all of them archaeologists belonging to 8 different public and private institutions in Spain (Incipit CSIC, Xunta de Galicia (the regional government of Galicia), private companies in the area of cultural management, the University of Santiago de Compostela, the University of Vigo and CCHS-CSIC).

Design Principles: As far as randomization is concerned, the objects were assigned randomly to the subjects. The tasks were carried out with a visualization (not selected randomly) identified for that task and with an unidentified random visualization for the task. Later, all the visualizations were shown to the participants so that they could choose their favorite one to carry out the task in question.

As far as the need for blocking variables is concerned, no variables were found to block, although the participants' prior skill with the visualizations which they were presented with was taken into account via an initial questionnaire. Last of all, the principle of balancing was applied manually to ensure that the selected visualization was evaluated by all the participants and that the rest of the visualizations were evaluated (tasks were performed with them) by a balanced number of participants in each case.

Instrumentation: The objects were not selected randomly due to the fact that the visualizations were carried out ad hoc for this empirical study. A list of the visualization techniques to be evaluated was made beforehand and a visualization was made with real data from archaeological data sets for each of the techniques (see Table 5.2):

Table 5.2 Information visualization techniques evaluated in the empirical study 3

Visualization techniques
1. Stacked bar chart
2. Line-based chart
3. Simple bar chart
4. Bubble-based chart
5. Customized Venn diagram
6. Treemap
7. Geographical map
8. Scatter chart
....

The tasks to be carried out were:

A. The participants had to respond about the specific number of archaeological objects present in the data which comply with several combinations of specific values: S shape and associated decoration and balloon shape without decoration.
B. The participants had to respond about the number of objects which make up the groups of decorated and non-decorated elements according to their shape. Then, they had to group them inversely, responding about the number of grouped elements by shape according to whether they were decorated or not.
C. The participants had to identify the increase or decrease in the groups of archaeological objects which were decorated according to their shape: Which shape was the most decorated? And the least?
D. The participants had to respond regarding the specific dates when the actions for the preservation of the objects were carried out and if there was a change in the registered state of conservation which could be put down to this intervention.
E. The participants had to identify the number of sites marked as sources in the data of the "Hillfort" type and "Iron Age" period in Galicia (region of Spain).

It should be noted that the tasks defined attempted to maintain a low level of difficulty due to the fact that many authors have stated that the complexity of tasks combined with a high degree of freedom can lead to a large number of errors, which would be reflected in our rate and would not be attributed to our dependent variables [155]. Figures from 5.9, 5.10, 5.11, 5.12, 5.13, 5.14, 5.15 and 5.16 show the visualizations created with archaeological real data for the carrying out of the study.

Fig. 5.9 Visualization 1: Stacked bar chart

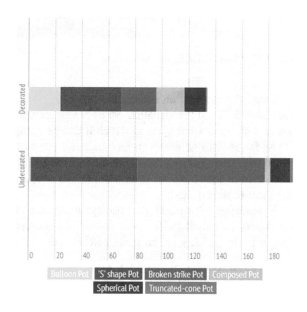

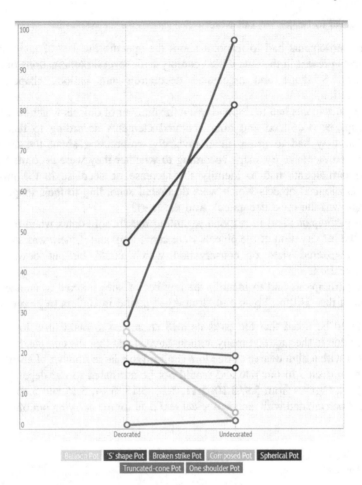

Fig. 5.10 Visualization 2: line-based chart

Later, the distribution among the tasks (with the associated cognitive process, see Table 5.3) and the visualization technique thought to be most appropriate for it, along with other techniques to which it can be compared, was carried out.

As far as lines of action in the study are concerned, the participants were provided with some instructions about the tasks to be carried out and the visualizations to be used in each case. No prior training was necessary for the participants in order for them to do the study.

Evaluation of aspects of validity: Following Wohlin's definition, based on Cook and Campbell [54], it was considered necessary to highlight the following threats to the validity of the study beforehand:

– As far as its internal validity is concerned, it was considered that the number of participants in the study minimized this risk as it was considered to be an

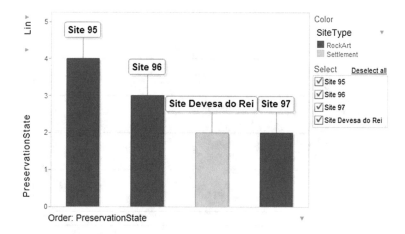

Fig. 5.11 Visualization 3: simple bar chart

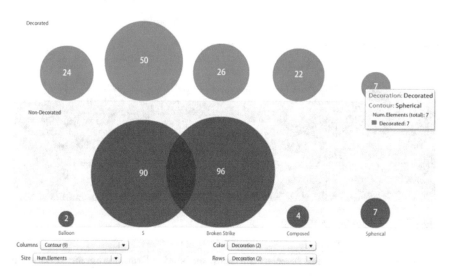

Fig. 5.12 Visualization 4: bubble-based chart

acceptable number. Furthermore, as was explained at the beginning of the chapter, the aim of this empirical study was not to establish these relations, thus this threat was minimized.

− As far as its external validity is concerned, it was considered that the probability of the results being repeated in other archaeological contexts is high, although its generalization to other sub-disciplines of heritage is a risk. All of this can be seen in other studies carried out in heritage [172].

− As far as its conclusion validity is concerned, it was considered necessary to choose objects which represent data from other origins in order to minimize the

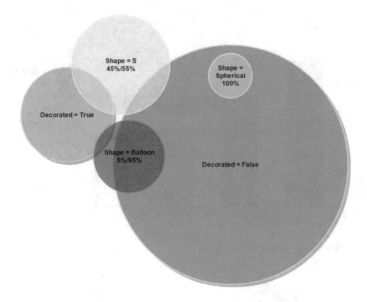

Fig. 5.13 Visualization 5: customized Venn-diagram

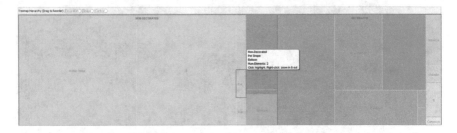

Fig. 5.14 Visualization 6: treemap

possibility of the data selected affecting the study. Conducting the sessions in person, with only one participant at a time, minimized the threats regarding the quality of the data obtained.

- As far as its construct validity is concerned, a possible threat could be the suitability of the selected measures. The differences in the participants as far as their skills with the visualization techniques to be evaluated are concerned could also have an influence when it comes to evaluating both methods.

The Execution of the Study—Operation

Preparation: The participants were not aware of the data or the visualizations employed. They were told about the characterization of the cognitive processes to be used but not about the hypotheses being dealt with in the study. By offering themselves as volunteers for the study, they gave their consent for this. The

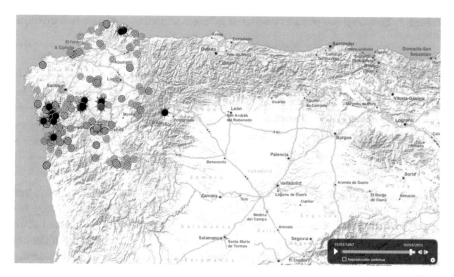

Fig. 5.15 Visualization 7: geographical map

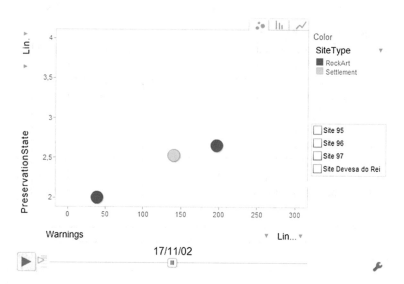

Fig. 5.16 Visualization 8: scatter chart

necessary materials were prepared beforehand: an initial questionnaire about the participants' professional and personal profiles and any previous experience with these types of visualizations, the statements for the tasks and the visualizations.

Execution: The participants were received one by one in individual sessions lasting between one hour and one and a half hours. They were asked to read the statements

Table 5.3 Task/Cognitive process visualization technique selected other visualization techniques compared

Task/Cognitive process	Visualization technique selected	Other visualization techniques compared
A/Combining	4 (Bubble-based chart)	1, 2, 3, 5, 8
B/Clustering	4 (Bubble-based chart)	1, 2, 3, 5, 6
C/Clustering-building	6 (Treemap)	2, 4, 8
D/Combining-building	8 (Scatter chart)	1, 2, 3, 4, 5
E/Situating	7 (Geographical map)	4, 8

and carry out the tasks. The visualizations stored their answers automatically, thus providing us with information about errors and correct answers. After they had carried out the tasks, they were shown all the visualizations and were asked about which of them they had used for each one of the four cognitive processes of the first level of abstraction of the proposed characterization (Combining, Clustering, Situating and Building).

The validation of the data: No invalid data was detected. This can be attributed to the individuality and the supervision of the process of the study's execution.

Analysis and Interpretation: Results
Results according to mistake rate per task and visualization MR (V_x):
Figure 5.17 shows the distribution of the mistake rate MR according to the visualization employed to carry out each of the tasks. As can be seen, the visualization with the lowest mistake rate in all cases is the selected visualization, except in the case in which the process being assisted is Clustering/Building. That is to say, except when we attempt to assist in a better understanding of the internal structure of information in order to form groups. In that case, the selected visualization is the Treemap but the mistake rate for other visualizations demonstrates better behavior in tasks associated to Clustering/Building than the TreeMap visualization. Therefore, hypothesis H_1 is confirmed except in this case, which enables us to detect the necessity for a better adapted visualization for this specific case than those proposed in the study.

   Results according to Distribution of Preference (DP): Figure 5.18 shows the distribution of preference DP. The participants' choices generally coincided with the visualizations determined beforehand as those best adapted to the cognitive process associated to each task, although no clear technique was observed in the case of the Building tasks. That is to say, the participants did not find it any simpler or more intuitive to access the structure of the information via one technique, with the technique designed specifically (the treemap) proving to be particularly complex. Due to this fact, we can consider establishing techniques better adapted for the visualization of the structure of the information as a line of action.

   Results according to the general mistake rate per task carried out (MRT):
Figure 5.19 shows the behavior of MRT. As can be seen, tasks A, B and E presented extremely high ratios of correct answers with very few errors. These tasks are those relating to combinations of values, grouping or other similar activities in

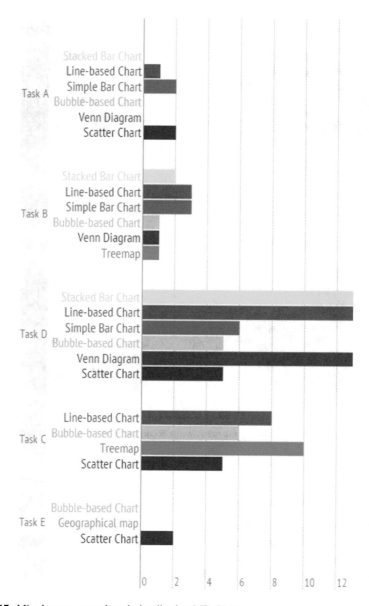

Fig. 5.17 Mistake rate per task and visualization MR (Vx)

geographical contexts. However, tasks C and D show much higher MRT values. These tasks, corresponding to the detection of tendencies and searching for correlations between temporal values, indicate that perhaps it is necessary to emphasize these points with the creation of more adapted visualizations than those presented here.

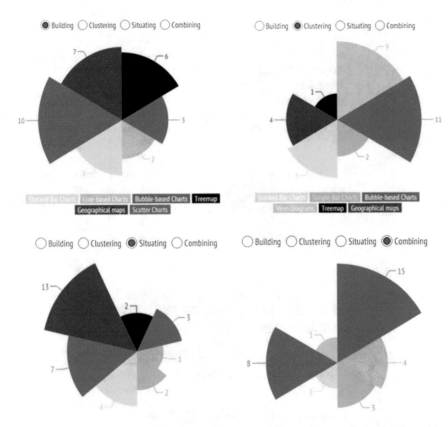

Fig. 5.18 Preferences expressed by the participants according to the type of associated cognitive process (DP)

Presentation and Package
This appendix represents the presentation format for empirical study 3.

Summary of Results and Conclusions
This empirical study enabled us to gain an initial idea of which mechanisms for the visualization and presentation of information provide better results, with a lower mistake rate, among archaeologists. In addition, qualitative information was gathered on the specialists' preferences regarding the visualizations.

In summary, the grouping tasks showed better results with mechanisms based on bubble charts, detecting the need for improvement in the choice of colours used (the participants looked for and/or attributed a meaning by colour) in the relative situation of groups on the screen.

In tasks relating to the analysis of the structure of the information, treemaps [248] were selected as the most suitable type of visualization. However, they presented bad results and the data of the mistake rate (MR) and the preference distribution (PD) corresponding to the rest of the visualizations used to infer the

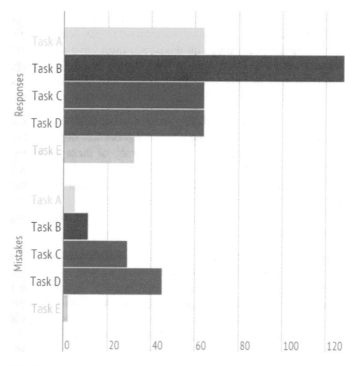

Fig. 5.19 Mistake rate compared to participants' answers for each task carried out (MRT)

structure of the information does not allow us to extrapolate a clear preference for any type. Work must be continued on visualization in order to provide support for this type of cognitive processes.

As far as the tasks relating to the combination of values are concerned, bubble charts gave the best results, although the specialists later expressed preferences for both bubble and scatter charts.

Last of all, context situation tasks presented good results with almost any type of visualization, due to the fact that they are dealt with as yet another combination value. However, the archaeologists requested a geographical map of spatial location in the context for this type of data, although it did not allow them to then carry out the rest of the tasks.

In addition to this information, the diversity of visualizations offered enabled us to extract some intuitive lines of action, the generalization of which is not possible due to the fact that they are not specifically instrumented in the study but they are of interest for future studies in this area:

• A good control of "*detail on demand*" systems was observed in all types of visualization, with search tasks relating to the structure of information providing better results.

- The archaeologists expressed quite a high degree of confusion regarding the meaning of positions on the screen, especially in categorization or grouping tasks: the position of the groups on the screen must have a semantic relation.
- The archaeologists expressed quite a high degree of confusion regarding the definition of phases in temporal visualizations: the visualization mechanisms shown did not prove sufficient to carry out the task without prior explanation.

In conclusion, we believe that these types of visualizations can provide good results when it comes to assisting the cognitive processes defined in our characterization, although it is necessary to deal with each of these cognitive processes individually and to work on several levels of abstraction with the mechanisms of presentation and interaction so that the proposed solution gains in flexibility and capacity for adaptation. Therefore, a more formal definition of the mechanisms for presentation and interaction will be necessary in order to be able to integrate them into our software assistance solution.

EMPIRICAL STUDY 4: The Empirical Validation of Visualization Techniques via the Use of TAP Protocols

The final empirical study carried out as far as presentation and software interaction are concerned, consisted again of the application of TAP protocols in an attempt to compare the perception of the participants, in terms of ease of use and understanding of aggregate information visualization techniques compared with their usual methods of data analysis (based on spreadsheets such as Microsoft Excel). In addition, an attempt was made to extract information about which factors are decisive in both techniques for the analysis of data, which visual aspects are noticed first in the analysis of data using both methods and which aspects are lacking in both methods which, in their opinion would enable them to analyze data more effectively.

In order to do this, data from a real archaeological case study was selected and a traditional method of analysis which was used to analyze the data was maintained (several Excel spreadsheets containing raw data accompanied by small pie charts serving to clarify the distribution of the data to the archaeologists). The same data was also presented in the eight types of visualization techniques being evaluated. The definition and the full results of the study are described below.

Objective(s) of the Empirical Study—Scoping

In Wohlin's terms, the objective of the empirical study is expressed as:

(a) Analyze two methods of data analysis for the purpose of evaluation regarding the degree of understanding and the influence on the data analysis carried out from the perspective of archaeologists studying data analysis processes in the context of public and private heritage institutions.

Planning the Study

Context: This empirical study was carried out partly with real tools (the data sets was real as was its visualization for analysis in Excel) and partly with "toy examples" (all the created visualizations). The empirical study was set in a specific, though broad, context (a sample of archaeologists belonging to both public and private institutions on a national scale). It can be considered that the context of the study is on-line as it was carried out in the professional context in which it would be used. It should be remembered here that the study was carried out using a methodology based on TAP (*Thinking Aloud Protocol*), as described in Chap. 4.

Formulation of the hypothesis: Two hypotheses were verified. They can be informally described as:

- The method of data analysis used does not have an influence on the degree of understanding of the data on the part of the participant. This is reflected in the participant's perception regarding the level of readability and understanding of both methods, which does not vary.
- The method of data analysis used does not have an influence on the type of data analysis carried out. This is reflected in that a change in the method does not lead to changes in the participant's focus of attention (what does the participant focus on initially to analyze the data?). Although it will not be formalized, we shall also see if no changes occur regarding the aspects which the participant claims are lacking in terms of data analysis.

Based on these informal hypotheses, we can now define H_0 and H_0' as null hypotheses and their alternative hypotheses as:

H_0: The participants demonstrated a similar level of readability and understanding (RUL) to method 2 (with the proposed visualizations) compared to working with Excel (method 1). It could even be the case that, within the TAP protocol, the participants may show that their RUL is independent of the method used.

H_0: RUL (m_2) = RUL (m_1)

Alternative Hypothesis H_1: RUL $(m_2) \neq$ RUL (m_1)

H_0': The set of x factors shown as Focus does not vary according to the method applied. H_0': FF (x, m_1) = FF (x, m_2)

Alternative Hypothesis H_1': The set of x factors shown as Focus varies according to the method applied. H_1': FF $(x, m_1) \neq$ FF (x, m_2). We have the impression that this variation is caused with aspects more related to the analysis of the data itself than with the presentation method, which could indicate that the step towards the generation of knowledge is taken more intuitively when using information visualization techniques than when using the Excel spreadsheet method.

Selection of variables: The archaeologists' skills in both methods of data analysis were selected as an independent variable. These were controlled via an initial questionnaire. The dependent variable in each case would be the degree of understanding, formalized by way of the RUL variable and the influence on the analysis of the data, formalized by the vector FF(x) applying each method.

Selection of subjects: The sample of participants corresponds to the *Simple random sampling* model. A sample of six archaeologists coming from 3 different public and private institutions (Incipit CSIC, Xunta de Galicia (the regional government of Galicia) and a private company specialized in the area of cultural management) was selected. They were selected randomly from among the 25 specialists who expressed an interest in participating in the study.

Design Principles: As far as randomization is concerned, the objects were not assigned randomly to the subjects. In other words, all the participants evaluated both methods. The participants evaluated the methods randomly, sometimes method 1 first and then method 2 or vice versa. However, we believe that the order of evaluation is not relevant.

As for the need to block variables, no variables were found to be blocked, although the influence of the participants' skills in both methods of analyzing data would be taken into account, being measured via an initial questionnaire.

Last of all, it would have been desirable to apply the balancing principle. However, due to the difficulty in finding a high volume of subjects, all of them evaluated both methods so balancing did not occur.

The type of design for the empirical study would be a factor with two treatments for each hypothesis.

Instrumentation: The objects were not selected randomly, due to the fact that the visualizations were made ad hoc for the purposes of this study. As far as lines of action in carrying out this study are concerned, the participants were provided with a statement with the instructions for taking part. No prior training was necessary for the participants to answer the questionnaire.

The evaluation of aspects of validity: According to Wohlin's definition, based on Cook and Campbell [54], it was considered necessary to highlight beforehand the following threats to the validity of the study:

– As far as its internal validity is concerned, the number of participants in the study is low, which could compromise the results if the objective of the study were to establish causal relations. However, as was explained at the beginning of the chapter, this empirical study did not aim to establish this type of relations, thus minimizing this threat.
– As far as its external validity is concerned, it is considered that the probability of the results being repeated in other contexts is high, due to the randomization of the objects used, as well as to the fact that other similar studies and experiments have been carried out in heritage contexts [172].
– As far as its conclusion validity is concerned, it is considered to be necessary to choose objects which represent data from other origins in order to minimize the possibility of the selected data influencing the study. The fact that the sessions were conducted in person, with only one participant at a time, minimized the threats regarding the quality of the data obtained.
– As far as its construct validity is concerned, it should be taken into account that the small sample of participants did not allow us to make a statistical generalization of the results. What is more, it would be necessary to carry out a greater number of similar studies and experiments in order to avoid the "*fishing and error rate*" problem, in other words, for this study to succeed in identifying a relationship between variables which, in reality, would be the suitability of the selected measures. However, carrying out studies using the TAP protocol

enables more data to be obtained in the future. The differences in the participants, as far as skills in the methods involved are concerned, may also have an
influence when it comes to evaluating both methods.

The Execution of the Study—Operation

Preparation: The participants were not aware of the data or the visualizations
employed. They were not informed not about the hypotheses being dealt with in the
study. By offering themselves as volunteers for the study, they gave their consent
for this. The necessary materials were prepared beforehand: an initial questionnaire
about the participants' professional profiles and any previous experience with data
analysis methods and the data presented in both methods.

Execution: The participants were received one by one in individual sessions lasting
between 45 min and one hour, thus avoiding problems deriving from fatigue in
TAP protocols. They were all informed that they were going to be recorded,
according to the TAP protocol described in Fig. 19 (see Chap. 4). Then, they were
asked to evaluate (Low, Average, High) the level of readability and understanding
of the data with both methods. The following step was to ask them freely about
what aspects they focused their attention on in order to understand the data and
analyze it. Last of all, they were asked about what aspects they considered were
lacking in the presentation method in order to be able to analyze the data better. All
of this was explained by the participants, who spoke out loud about the reasons for
their choices and/or doubts.

The validation of the data: No invalid data was detected. This can be attributed to
the individuality and the supervision of the process of the study's execution.

Analysis and Interpretation: Results

Results according to Readability and Ease of Understanding (RUL):
Figure 5.20 shows the behavior of the values given by the participants regarding
readability and level of understanding on a scale of three values (Low, Medium and
High). As can be seen, the level of readability and the level of understanding both
increase with method 2, which corresponds to the use of visualization techniques.

   Of course, with this we do not claim to declare an improvement in the related
tasks but just an increase in perception regarding readability and understanding of
the data on the part of the participants. However, this aspect is of vital importance in
the research in question, given that assistance to the generation of knowledge must
also deal with perceptive aspects of the stakeholders themselves in the process.

Results according to the set of x factors of Focus of Attention (FF):
Figure 5.21 shows a spider chart representing several factors which have an
influence on a certain element, in this case those which the participants mentioned
in the *Thinking Aloud* sessions as relevant when starting their data analysis process.
As can be seen, the patterns of method 1 and method 2 differ significantly, thus
confirming our hypothesis H_1'. In spite of not having formalized this in a
hypothesis, we also wished to evaluate the behavior of the vector of factors referenced by the participants as aspects which they consider to be lacking when it

Readability & Understanding level

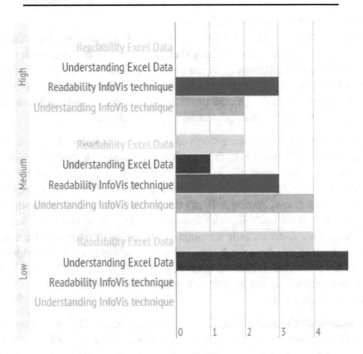

Fig. 5.20 Level of readability and understanding (RUL) expressed by the participants according to method of data analysis (Excel vs. information visualization techniques)

comes to analyzing data using both methods. This analysis presents also important variations depending on the analysis method employed –see Fig. 5.22—. In both figures, the factors on which they focus their attention when using the Excel data analysis method are shown in blue. The same analysis when using information visualization techniques is shown in orange.

Presentation and Package
This section represents the presentation format for empirical study 4.

Summary of the Results and Conclusions
This empirical study enabled us to gain an initial idea about the perception of archaeologists regarding readability and ease of understanding in both methods (a traditional method using Excel and that based on aggregate information visualization techniques). In general, the visualization techniques were evaluated more positively that the traditional method, both in terms of readability and ease of understanding, except in some specific visualizations which must be reviewed with a view to the general proposal of a software assistance solution. Furthermore, the study enabled us to extract data on how the behavior of the user changes when he/she analyses data

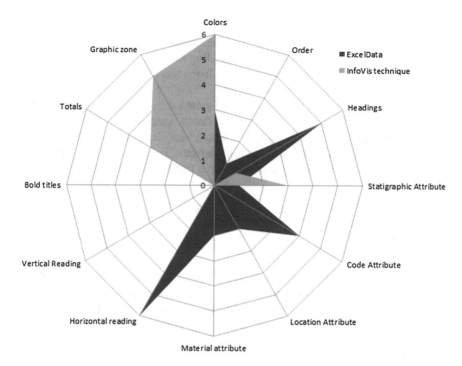

Fig. 5.21 Factors which attract the participant's focus of attention (FF) when using both methods of data analysis

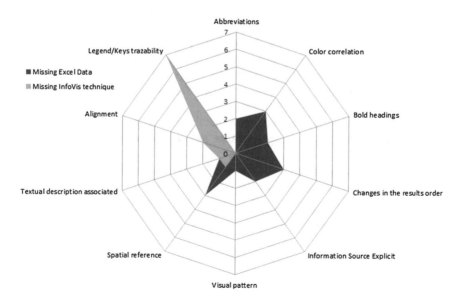

Fig. 5.22 Aspects sought but not found by the participants employing both methods of data analysis

with visualization techniques instead of using the Excel spreadsheet method. It was possible to verify that their behavior varied considerably: the participant pays more attention when analyzing data on aspects relating to more global analysis and less when using the visualizations offered to them. In the same way, the aspects which the archaeologist claim are lacking differ when changing method, with the use of visualizations proving to be more abstract and aimed at the generation of knowledge. All of these details allow us to corroborate software assistance to the generation of knowledge through offering this type of adapted visualizations as a line of action.

Both empirical studies 3 and 4 reaffirm our intuitive idea for the need to deal with aspects of interaction and presentation on a formal level in software assistance in archaeology.

Conclusions

The empirical studies carried out have enabled us to confirm intuitively, though not in terms of statistical representation, the importance of expressly taking cognitive processes and their support into account in terms of how archaeological data sets are visualized, with the aim of improving the knowledge generation process in the discipline by way of software assistance. More specifically, the software assistance we provide should:

- Respond to the most common cognitive processes in archaeology using the proposed hybrid characterization which links Hobbs' coherence relations, originating from archaeological discourse analysis, with our own level of cognitive processes defined for assistance in 4 keys and/or directions depending on the objective of the archaeologist: Building, Clustering, Situating, Combining.
- Integrate the temporal and geographical perspective in the assistance itself as an archaeological key data aspect but not as a separate cognitive process. It was proved that reasoning in terms of space and time was integrated into the presented typology of cognitive processes.
- Design mechanisms for presentation/interaction based on aggregate information visualization techniques, adapting these techniques to the cognitive process which we wish to assist.
- Establish mechanisms which allow us to monitor the software assistance offered in each case, integrating its different perspectives.

The large amount of existing works and empirical results collected until now evidences that, to provide a complete software assistance to the generation of knowledge in archaeology, it is essential to deal with all the elements involved (the conceptualization of data, cognitive processes to be assisted and mechanisms of interaction and presentation identified as being the most suitable) with integrally and on a conceptual level. This has led us to opt for the development of a framework as a solution, which will be described in the following part, enabling us to deal with all these aspects relating to software-assisted knowledge generation.

Part III
Software Assistance Strategies for Archaeological Data

Part III
Software and Statistical Methods for
Archaeological Data

Chapter 6
Framework Overview

Introduction

Over the course of the previous chapters, we have explored the co-research environment in software engineering and archaeology, clarifying why it is necessary to treat the research problems involved here from this co-research perspective and we have shown some examples of succeed cases. In addition, we have defined the initial hypothesis and research questions that we will asked along this book, studying how we can assist through software the knowledge generation process in archaeology. The work with archaeological data presents some characteristics analyzed also in previous parts that we must consider in order to model the adequate software assistance strategies. With all this in mind, throughout the Part III we combine this whole environment to propose to the reader how to model software solutions for assisting archaeologists and professionals' practitioners working with archaeological data to generate new knowledge from their large data sets. All strategies are articulated in what we call a conceptual framework.

By conceptual framework we understand an analytical tool [247] used in order to capture conceptual structures and distinctions between their elements, as well as to organize, communicate and prescribe these structures. Such a tool should be intuitive and easy to recall.

The use of conceptual frameworks in order to express proposals and solutions is widespread in the field of research [141], both in the contexts of deductive or inductive research methodologies and in different disciplines, such as in business [143], natural sciences and engineering [15, 151], humanities and social sciences [8, 161, 228, 237]. According to Shields and Rangarajan [247] a conceptual framework is "the way ideas are organized to achieve a research project's purpose."

In our case, the conceptual framework takes the shape of a set of conceptual models that represent the differing perspectives to be taken into account as far as software-assisted knowledge generation in Archaeology is concerned. This complete proposal will allow the reader to define and decide, step by step, the assistance

© Springer International Publishing AG 2018
P. Martin-Rodilla, *Digging into Software Knowledge Generation in Cultural Heritage*, Modeling and Optimization in Science and Technologies 11, https://doi.org/10.1007/978-3-319-69188-6_6

strategies that he/she needs to use with a particular archaeological data set, always maintaining as final objective the generation of knowledge inside each real archaeological case study.

Next section will present the structure and general characteristics of the proposed framework. In subsequent chapters, each part of the framework will be described in detail, illustrated by archaeological examples and discussed separately.

Modeling Software Assistance Strategies: Subject Matter, Cognitive Processes and Presentation and Interaction Mechanisms

The General Structure

As we explored in previous parts, we have found three complementary perspectives to examine in order to include them in a software assistance framework: subject matter, cognitive processes and presentation and interaction mechanisms.

As far as the first part is concerned, as will be detailed throughout Chap. 7, Archaeology specialists deal with a large quantity of raw data, that is to say, primary heritage data in need of analysis, which forms the basis of knowledge generation in the field. In other words, Archaeology and related sub-disciplines constitutes the subject matter on which new knowledge will be generated. As other authors have already identified [106], there are specific needs as far as the conceptualization and treatment of heritage information in concerned. The proposed framework adopts the *Cultural Heritage Abstract Reference Model* (CHARM) [99] as its abstract model of reference, which provides a solution to a large part of the problems detected in these studies. It should be noted that, although the framework adopts CHARM as a solution for this part of the framework, it is necessary to extend it for each specific implementation, due to CHARM's nature as a reference model. Chapter 7 details how conceptualize large archaeological data sets and prepare them for software-assistance knowledge generation process and applications.

As far as the second part is concerned, as will be detailed in Chap. 8, the generation of knowledge is, by definition, an ensemble of cognitive processes which human beings carry out based on some kind of data. As has been seen in Part II, approaches from the fields of Software Engineering and Archaeology to this cognitive dimension in software frameworks are scarce, especially as far as those relating to the generation of knowledge are concerned. However, cognitive processes are an intrinsic element which must be modelled if we wish to assist to knowledge generation, given that the carrying out of these processes is what enables a rise in the DIKW hierarchy of levels of knowledge generation [7], as they form an intrinsic part of them. The proposed solution consists of models which allow these cognitive processes to be incorporated into the framework and, therefore, into the software systems which are designed based on the aforementioned framework.

As far as the third part is concerned, as will be described in Chap. 9, we believe that the model of software assistance which fits best in the field of Archaeology should implement this assistance in the form of heritage information visualizations adapted to the theme being dealt with and to the cognitive processes which the user performs on them, according to Chen's model [61]. In order to do this, it is necessary for the framework to incorporate software presentation and interaction models which define how this assistance is implemented via visualization in each case. Other types of assistance, which were considered initially, such as the application of knowledge extraction techniques, have been taken into account in the definition of the framework and have been implemented into it by way of iteration mechanisms, although they have not been completely developed by archaeological case studies in this book. However, these other types pf assistance form part of future research and we can discuss it at final chapters.

We have now an overview of the different parts that we must model if we want to prepare and develop our archaeological data for software assistance to the knowledge generation. Later chapters shall guide the reader to the different parts of the framework, their software models produced in archaeological cases and the scientific contributions which they present.

Chapter 7
Archaeological Subject Matter

Conceptual Modeling for Archaeological Data Sets

An adequate conceptualization of the domain involved is the first step in any software technique, process, tool or system that we would want to build. In our case, it becomes necessary to characterize the nature and specific characteristics of the archaeological data used as a source by the specialists for the generation of knowledge. As has already been explained previously, conceptual modeling proves an appropriate technique for a definition on a conceptual level of the sphere of application or the subject matter of the framework here proposed.

Conceptual modeling applied to archaeological data has gained special relevance in recent years, as has been described in Chap. 4. However, the need for a more engineering-based focus and the fact that existing solutions lack a mechanism to resolve the tension identified between prescriptive models and their necessary personalization has led to the appearance of modeling approaches from a formal, but integrative, perspective such as CHARM [99] or the language in which it is expressed (ConML) [96]. Both projects constitute the conceptual basis from an archaeological subject point of view of the framework proposed.

CHARM (Cultural Heritage Abstract Reference Model)

The Cultural Heritage Abstract Reference Model (CHARM) is a semi-formal representation model of Cultural Heritage on a conceptual level. As a reference model, it is broad and shallow, aiming to cover as far as possible the social and cultural phenomenon which we know as Cultural Heritage, albeit at a high level of abstraction. This allows it to be used by a wide range of users and provides it with a great degree of flexibility for a wide range of objectives and purposes [99].

© Springer International Publishing AG 2018
P. Martin-Rodilla, *Digging into Software Knowledge Generation in Cultural Heritage*, Modeling and Optimization in Science and Technologies 11, https://doi.org/10.1007/978-3-319-69188-6_7

CHARM model takes a Cultural Heritage definition embracing a wide range of topics and visions of heritage, as we referred in Chap. 1. Although we are adopting the same definition along this work, our main purpose at this part of the framework is organizing archaeological information models conceptually, in order to subsequently defined software assistance mechanisms for each specific case study. Thus, next sections of this chapter will explain CHARM in general as a model for Cultural Heritage domain and subject (as CHARM's authors definition), although our conceptual models and resulting data structures for the case studies presented in this book are oriented only to archaeological information and their applications.

Below, we shall describe CHARM's approach and its most relevant characteristics for this research and what specific contributions have been made to it, deriving from its use as a foundation for the proposed framework.

The Characteristics of CHARM

CHARM is expressed in ConML [96, 97], a conceptual modeling language which broadens the conventional focus oriented towards objects with its own characteristics:

- ConML is specifically designed to be used by people who are not experts in the fields of Information Technology and Conceptual Modeling. Therefore, it attempts to eliminate complexity, technical vocabulary and the steep learning curve required by non-experts in Information Technology when it comes to modeling with other existing languages such as UML [203]. By using ConML, archaeologists, Cultural Heritage researchers and related practitioners are able to create a conceptual model of their objects of study, draw up a data and/or information model based on it and guarantee a conceptual connection and interoperability with other existing CHARM models.
- ConML incorporates support for "soft aspects" of modeling, such as temporality and subjectivity, which are not supported by other modeling languages which have a more general purpose, and which are especially relevant in archaeological modeling [96]. This means that CHARM can incorporate the archaeological basis as well as related disciplines needed in order to model the complexity reality of any archaeological case study and their data involved.

CHARM is made up of 163 classes [127] in version 0.8 and 173 classes in version 0.9 [130]. This research uses CHARM 0.8, although the stratigraphic information is based on version 0.9, as detailed below. CHARM describe 3 fundamental areas of Cultural Heritage, in which any entity is susceptible to being defined within Cultural Heritage, receiving cultural value from individuals or communities. Therefore, CHARM allows an entity to be documented apart from the different evaluations which individuals and/or communities may make of it. Thus, within CHARM we find an area entitled "Evaluable Entities" and another named "Valorizations". In the end, both kinds are able to be captured by way of representation, which delimits a third fundamental area within CHARM: "Representations". If we go deeper into the conceptual sphere of the three areas, they can be defined as:

- Evaluable entities: those entities belonging to reality which have received, receive or are susceptible to receiving cultural value from individuals and/or communities. They are, therefore, entities from which heritage is built socially, such as a building, a song, an archaeological site, a painting, etc.
- Valorizations: those entities of a narrative and/or discursive nature which add cultural value to an evaluable entity, generally by way of subjective interpretation mechanisms by a well-defined individual and/or community. They represent, therefore, the added cultural value which converts an evaluable entity into a heritage entity. Good examples of this may include technical reports on the historical importance of a building or a work of art or a manifestation of attachment to a particular place on the part of an association of neighbors.
- Representations: those entities which capture characteristics or properties of other evaluable entities (known as "content") and reflect them onto another evaluable entity (known as "embodiment"). Thus, these entities become representations of existing evaluable entities, such as a painting, a photograph or a 3D model of a building.

In addition to these three areas, the CHARM model also includes other aspects which are especially relevant for Cultural Heritage (and subsequently for archaeology), such as geographical locations, temporal aspects, processes, measurements and dimensions and agents which define, interact, modify or take decisions regarding heritage, among others. This allows those cross-cutting aspects to be included in any of the three previously defined areas. Figures 7.1, 7.2 and 7.3 show the three areas which have been explained. For more detail on each of the classes, as well as on other cross-cutting aspects dealt with in CHARM, see [130].

In brief, CHARM expresses concepts relating to real entities which may receive heritage value, what type of evaluation each entity receives and if a specific entity is represented in, or represents, others. In addition, it expresses where the entities are located, which of their temporal aspects and processes we wish to deal with, how we conceptualize their dimensions and what agents interact with them. Therefore, it offers the necessary conceptual support to express in the archaeological subject area which the reader is dealing with in each case and upon which he/she is generating new knowledge. However, CHARM is an abstract reference model, which is to say that it is not conceived to be used by itself for practical objectives. Rather, it needs to be extended and adapted for each case in which it is applied. In the following section, we shall introduce CHARM's extension mechanisms and describe how they have been used in the proposed framework.

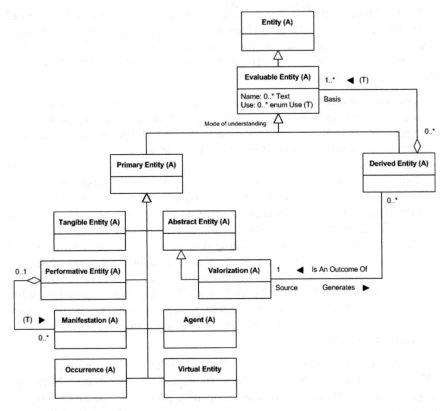

Fig. 7.1 Evaluable entities and their hierarchy of classes in CHARM model, expressed in ConML language

CHARM's Extension Mechanisms

By extending CHARM, its scope is reduced and its depth and precision in the level of abstraction are increased, so that the archaeological subject area or the project being dealt with can be perfectly described. Both the ontological and epistemic reasons which determine the suitability of a model for a particular situation are complex, as has previously been mentioned [98, 215]. Therefore, the extension of CHARM always implies the production of a particular model, a CHARM super-set, so that each class in the particular model which is not in CHARM is compatible in Liskov's terms [167] with a CHARM class. Being compatible in Liskov's terms means that each entity represented in the particular model is also represented by CHARM and, therefore, can also be dealt with, from a more abstract point of view, as the corresponding class in CHARM. This is achieved via the specialization mechanism within the Object Oriented Paradigm [32].

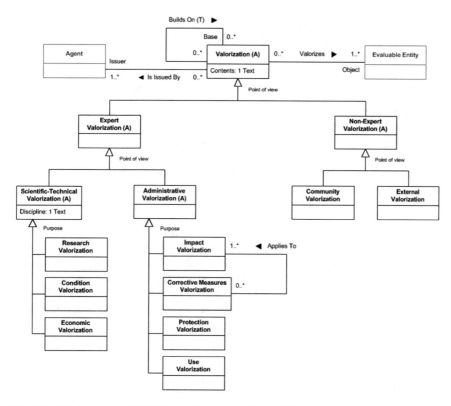

Fig. 7.2 Valorizations in CHARM model, expressed in ConML language

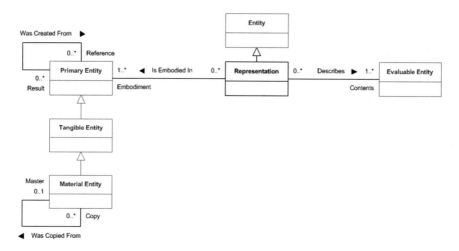

Fig. 7.3 Representations in CHARM model, expressed in ConML language

When a particular model is created, it is initially equivalent to CHARM. Later, CHARM extension mechanisms can be used, based on those provided by ConML, which can be combined when necessary and which allow the user to achieve a particular model which is closely adjusted to his/her personal purpose. CHARM/ConML extension mechanisms include actions such as:

- The addition of classes to the particular model by way of the specialization of one or more existing classes in CHARM.
- The elimination of CHARM classes which are not connected with other classes with a minimum cardinality greater than zero.
- The addition of attributes to CHARM classes in the particular model.
- The addition of associations between the classes in the particular model.
- The addition of enumerated types and elements to the particular model.

These extension mechanisms offer us a great degree of flexibility due to the fact that, by adopting CHARM as a basis for expressing the subject conceptual model of the framework, they allow the framework to always be able to handle a personalized subject model for the archaeological area which the reader is studying, whilst preserving the advantages of having CHARM as a reference model [128]. Another example developed in the course of this book, among others, regarding these advantages can be consulted in [105].

Other Structural Aspects in CHARM

CHARM, like any model expressed in ConML [129], has different modeling mechanisms at its disposal, provided by ConML's metamodel, which lends it flexibility when it comes to representing perspectives of heritage reality. The adoption of CHARM for this research mainly takes into account two modeling mechanisms which take on special relevance in the proposed framework and which are provided by ConML's metamodel [129]: the package mechanism and the cluster mechanism. Due to their use as mechanisms to conceptualize our archaeological data sets following the proposed framework, both structural mechanisms are described below.

Packages

According to ConML's technical specification, a package is a group of related classes, enumerated types and possibly sub-packages [129]. Figure 7.4 shows the part of the ConML metamodel in which the package mechanism is specified.

It should be noted that CHARM's official specification does not determine any default packages in the model. This, according to the specification of ConML [129],

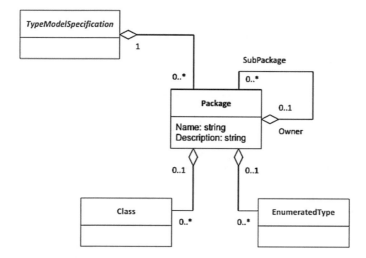

Fig. 7.4 Package metamodel fragment of ConML 1.4.4. technical specification

allows the user of a model expressed in ConML, such as CHARM, to define his/her own criteria in the definition of packages in the extensions which he/she carries out when working with CHARM. In our case, this mechanism constitutes a conceptual tool to delimit the subject sphere (a sphere is defined by a package) upon which the reader is to generate knowledge and, therefore, the subject sphere to be assisted by the framework here presented.

Clusters

According to ConML's technical specification, a cluster is a group of tightly related classes that usually work as a whole under a particular view of the target type model. A cluster always has a main class, which determines most of its semantics, plus a collection of participant classes. For this reason, the cluster name coincides with the main class's name. Clusters may be related between themselves. One instance of a Cluster is named Bundle in ConML [129]. Figure 7.5 shows the part of the ConML metamodel in which the cluster mechanism and its corresponding structure in the model of instances are defined.

It should be noted that CHARM's official specification does not determine any default clusters in the model. This, according to the specification of ConML [129], allows a user of a model expressed in ConML, such as CHARM, to define his/her own criteria in the definition of clusters in the extensions which he/she carries out when working with CHARM. In our case, this mechanism constitutes a conceptual tool to delimit sub-areas of the archaeological subject with structural and semantic features which are of relevance for the framework proposed.

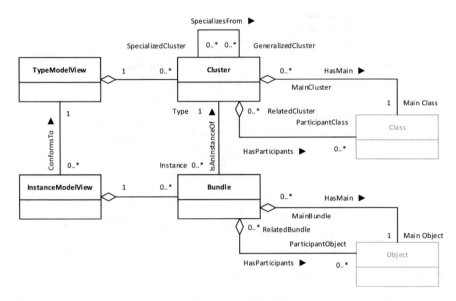

Fig. 7.5 Model views structure in the metamodel of ConML 1.4.4. technical specification

The package and cluster mechanisms work independently. Packages are conceived with the aim of providing a subject class structure to a model, whereas clusters form a view of the model for a specific aim. This allows clusters to work as a layer which is superimposed on the model organized into packages, overlapping several of them. In the following section, we shall describe the complete proposal for the integration of CHARM and the mechanisms described above into the proposed framework.

Our Proposal: CHARM as the Basis for the Knowledge Generation

We propose the use of CHARM as a reference model and its consequent extension to reflect, in each case, the archaeological subject matter about which knowledge is generated. Furthermore, the proposal is completed with the integration into the framework of the particular models of CHARM which are created. This integration requires the use of the modeling mechanisms provided by the ConML metamodel [129] (packages and clusters) explained above, with the aim of delimiting the subject spheres and their subsequent treatment in the proposed assistance to the generation of knowledge.

The Use of the Package Mechanism

The package mechanism provided by the ConML metamodel [129] shall be used in the framework in order to delimit sub-areas of the archaeological subject within the particular model being worked with in each case of software assistance. This particular model (an extension of CHARM) will be able to be subdivided into as many packages as necessary, which can fit inside each other, according to the defined ConML metamodel. For example, an extension of CHARM for a specific project will have as many packages as sub-areas of the subject contemplated in the project. The name of the package should correspond to the main subject concept of each selected sub-area. In other words, it should correspond to the class with the greatest semantic load within the package.

The Use of the Cluster Mechanism

The cluster mechanism enables us to identify sets of classes with their own semantics. In this way, we will be able to know which classes are relevant within the model and which of them have an auxiliary nature. The latter will be dealt with in the framework as characteristics or features which are specific to the main class of the cluster. All of this information will be defined in the interoperability model defined also as part of the framework (see Chap. 10).

Clusters can be defined according to the following criteria:

- The definition of clusters formed by at least one participating class, as well as the main class, is recommended. The structure of clusters formed by only one main class is not taken into account when later assigning them a behavior in the framework.
- The main class of each cluster brings together its main semantics, whereas the participants will act as auxiliary characteristics to the main one.
- The cluster is defined at the highest level of abstraction possible at which we wish to work in the model, inheriting affiliation to the cluster by all the specialized classes of each of the main and participating classes which were originally defined in the cluster.

This proposal for the use of the package and cluster mechanisms existing in ConML aimed at software assistance to knowledge generation is exemplified and validated on an analytical level via the real case study of A Romea, which has been selected for illustration along this book. The resulting models (instances of this proposal) for A Romea are implemented in a functional prototype. Both processes are detailed in Part IV.

Conclusions

As a result of what has been seen in this chapter, the reasons for the choice of CHARM (and, therefore, ConML) as the conceptual and modeling basis of the archaeological subject matter in the framework for assistance to the generation of knowledge presented can be summarized as follows:

- The modeling language employed is specifically designed for specialists in Humanities and Social Sciences (in which we placed archaeological data projects) and supports the specific characteristics of these fields.
- The extension mechanisms it has allow us to achieve a high level of personalization as far as subject matter is concerned in the particular models which are created. This personalization is necessary as software-assisted knowledge generation depends on the subject sphere of the field in question in the framework. However, the particular models maintain a common semantic and structural reference via CHARM, which allows them to obtain high degrees of compatibility and interoperability, as has previously been explained.
- The package and cluster mechanisms it has have allowed us to propose their use within the context of the framework, which is especially appropriate in order to be able to incorporate structural information into the interoperability model which is to be defined later and which establishes a bridge on a conceptual level between the remaining parts of the framework. This bridge will be seen in more detail in Chap. 10.

The use of the package and cluster mechanisms described in this chapter is part of the original contributions of this book. These mechanisms allow the reader to carry out personalized characterizations of subject aspects of the particular archaeological data models for their case studies, in order to then connect them through the interoperability model of the framework with the rest of the dimensions dealt with (cognitive processes and software presentation and interaction mechanisms). To conclude, it should be highlighted that this first part of the framework and its application to the archaeological case study carried out in subsequent chapters serve as a complete validation and illustration of the CHARM model itself as mechanism to represent archaeological data models and ontologies.

Chapter 8
Cognitive Processes

Introduction

The previous chapter described the conceptual support for the expression of all the archaeological subject aspects for a given case study using the framework proposed.

As was explained in previous parts, one of the aspects detected over the course of this research as being fundamental to software-assisted knowledge generation is the specific treatment within the framework of the cognitive processes which the domain specialist carries out on the previously defined subject model. Therefore, during the phase of exploring the problem, a characterization of cognitive processes in archaeology was built up, allowing us to express which processes are more relevant when it comes to designing a possible system of software assistance to the generation of knowledge in the field. In the same way, a methodology was constructed which allowed for the integration of discourse analysis into software engineering processes in order to permit the extraction of cognitive processes from textual sources, a common source of archaeological information. It should be noted that, thanks to all this prior work, these cognitive processes were revised, studied and described in depth, as can be seen in Part I. This chapter directly presents our way for dealing with cognitive processes in the proposed framework, referring to Part I as the investigative context of this work.

Expressing Archaeological Cognitive Processes

The part of cognitive processes in our framework consists of a basic metamodel with two main classes: *Coherence Relation* and *Inference Type*. A coherence relation is a connection between two discourse elements characterized according to a functional criterion. In other words, the types of coherence relations attempt to characterize what relations exist between units of discourse, mainly between clauses

© Springer International Publishing AG 2018
P. Martin-Rodilla, *Digging into Software Knowledge Generation in Cultural Heritage*, Modeling and Optimization in Science and Technologies 11,
https://doi.org/10.1007/978-3-319-69188-6_8

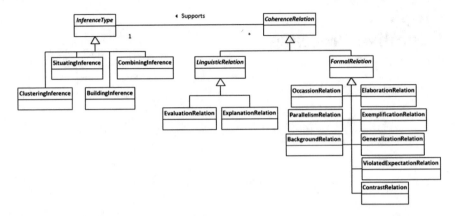

Fig. 8.1 Metamodel expressed in UML representing the cognitive processes in the framework proposed

and/or sentences [120] in the discourse and when each of them is used. On the other hand, an inference type is a set of coherence relations which maintain a common objective on the part of the archaeological specialist when carrying out the specific cognitive process.

As can be seen in Fig. 8.1, one or several coherence relations can contain one or several inference types, depending on the structure of the coherence relation in the discourse. In the metamodel, the coherence relations maintain the classification in linguistic relations and formal relations (on the basis of the degree of formalization obtained) as has been described in Part II. The formalization of the ten types of coherence relations identified by Hobbs, with each type of coherence relation constituting a sub-class of *Linguistic Relation* or *Formal Relation* in the metamodel, also maintains the structure described in Part II.

In the same way, each of the inference types identified in the characterization of cognitive processes carried out constitutes a specialized sub-class of *Inference Type*, in response to the definitions developed during Part II. Figure 8.1 shows the full metamodel of cognitive processes which is used in the framework. It should be noted that, unlike the models developed for the subject part of the framework, the cognitive processes metamodel works with classes which conceptualize "types". That is to say, the classes present in this metamodel conceptualize types of inferences and types of coherence relations and not specific inferences or coherence relations (with the latter being instances of these classes). We are, therefore, working on a higher level of abstraction than in the subject part. This is due to the fact that the framework must specify the subject part on the level of the discipline so it needs to go down to a more specific level of abstraction in order to achieve this while working on the level of inference types (without taking into account more specific concepts) when it conceptualizes cognitive processes. This allows this model to be used in order to relate the discourse analysis carried out with the inference types which have been characterized, using our abstract characterization for archaeology for other disciplines, or even adding more inference types which could be identified in the future.

Conclusions

This chapter presents the metamodel of cognitive processes developed for the framework and characterizes within it the inference types which have been identified as most common and necessary in order to assist to the generation of knowledge in archaeology and its connection with the coherence relations which are present during discourse analysis. This model allows us to work with the cognitive processes which the archaeologist habitually carries out. Furthermore, if the methodology presented in Part II for discourse analysis has been previously applied, it allows us to state in the framework which coherence relations identified during the discourse analysis support which inference type, thus enriching our knowledge regarding the cognitive processes to be assisted which arise in each case of application of the framework and allowing the framework to respond according to this information.

As has been emphasized before, the explicit incorporation of software models which include cognitive processes in the framework in order to assist to the generation of knowledge is one of the main innovative contributions of this research. The connection in the metamodel of these processes with the results obtained from the application of the methodology for discourse analysis presented allows us to maintain traceability from the greatest source of information regarding how knowledge is generated in archaeology: textual sources. The step by step modeling, the resulting models and some application examples for expressing archaeological cognitive processes in a specific case study with real archaeological data will be explain in detail in Part IV.

Chapter 9
Presentation and Interaction Mechanisms

Introduction

In recent decades, the strategic importance of software data interaction and presentation techniques for the analysis of large volumes of data (Big Data), along with their use in decision making, has grown considerably with the appearance of emerging disciplines [148], professions [66] and techniques which assist human beings in the handling and interpretation of data. Decisions based on large volumes of data are taken on a daily basis in numerous disciplines, such as decision making in business (BPM—Business Process Management) [65, 250], qualitative research, market research, statistical reasoning, etc. Archaeological data are not an isolated case and also require data analysis techniques which provide assistance to the researcher as far as the interpretation of and decision making based on that data is concerned, in knowledge generation in the discipline. See, for example, how this need is reflected in the empirical results regarding types of visualizations with archaeologists, as well as in interviews and bed tests regarding interaction and presentation mechanisms, all of which are described in Chap. 5.

Due to this need, the framework for assistance presented in this book should deal conscientiously with the formal representation mechanisms of data presentation and interaction which exist, in order to later evaluate them and identify the specific needs in archaeology. Then, it is necessary propose a solution which is correctly integrated into the dynamics of the data and reasoning processes which have previously been defined. This chapter presents an analysis of formal interaction and presentation techniques which are currently in existence, in what applications they are used and which disciplines they represent. This first broad tour for the current situation could be less interesting for practitioners and readers who wants to apply the third part of the framework directly, while represents a start point for researchers in the area. Following sections continue with the framework presentation, identifying a series of problems which are independent of the field of application being dealt with, and how we can solve them through interaction strategies using our

© Springer International Publishing AG 2018
P. Martin-Rodilla, *Digging into Software Knowledge Generation in Cultural Heritage*, Modeling and Optimization in Science and Technologies 11, https://doi.org/10.1007/978-3-319-69188-6_9

framework. The interaction mechanisms, based on design patterns, will be explained in detail in an abstract way in order to then describe its application to archaeological data sets.

Formal Representation in Software Presentation and Interaction for Data Analysis

The growing demand for software systems which allow for the extraction of data and assistance to decision making processes based on that data has recently encouraged research into the creation of mechanisms which enable the formal representation of software presentation and interaction with data. One of the areas experiencing the biggest expansion at the present time is that of InfoVis or Information Visualization [21, 40]. This emerging field draws together knowledge and skills from diverse disciplines (Psychology, Graphic Design, Data Analysis, Computation, Software Engineering, etc.) in order to create solutions for the visualization of data which allow for greater understanding and assistance to the user in the analysis of the data. These visualizations are widely used for a wide variety of purposes, among which we can highlight Infographics [38] and the field of Education, which have been enhanced enormously by *MOOC* (*Massive Open Online Course*) and other types of online courses [1].

The majority of visualizations which arose when InfoVis first appeared as a discipline were generated on demand [14] in order to represent a specific dataset (a collection of related data) and to facilitate its understanding and use, or were based on specific visualization techniques such as tree maps [248]. These ad hoc solutions could be immediately adjusted to the data which was to be visualized, providing the user with a satisfactory experience and fulfilling the task of providing decision making assistance. However, InfoVis researchers soon identified the fact that these visualizations did not allow for the reuse of the specific solutions for formal representation and interaction which they implemented [38].

Later, studies appeared regarding the definition, design and evaluation of specific visualization techniques, as well as presentations according to the type of data being visualized or the audience at which the visualization was aimed. Some of these approaches are at work in the present with abstract specifications of the visualization techniques and the formal representation of the interaction, leading to abstract code libraries [33, 229, 238]. These libraries allow for the reuse of the solution with different datasets to be visualized but they deal with its formal specification and its specific implementation as a whole. This situation supposes the limitation of the capacity of reuse and formal representation of the components of the solution, due to the fact that the underlying conceptual model, which the adopted interaction solution represents, is excessively connected with the specific implementation chosen in the specific case.

Recently, new efforts have been made to create languages which facilitate the formal definition of software interfaces for the visualization of data, in an attempt to uncouple the connection between implementation and the formal definition explained previously as [177, 179]. Another example is the language created by IBM [269] to define and specify interaction solutions in an abstract way, thus allowing for their reuse. In spite of the inherent advantages of the new approach, this solution still requires specific modeling skills on the part of the software analyst, which reduces its degree of applicability, especially due to the significant learning curve which it presents. Specification languages are, therefore, an integral solution with numerous advantages in large software development teams and companies but are difficult to apply in other contexts.

Both approaches (abstract code libraries and specification languages) are currently being applied for the formal representation of presentation and interaction mechanisms with data [196], with the advantages and disadvantages explained above. However, the problems regarding the formal definition of presentation and interaction mechanisms have also been dealt with from another perspective by software engineering, highlighting the proposals focused on the use of interaction patterns.

The use of patterns consists of the repetition of a previously applied solution to similar problems to that which we wish to solve, independently of the field of application and the skills of the software analyst or specialist in the field who designs the interface [10]. There is a broad corpus of research regarding the modeling of interaction based on patterns, from those based on the elements of the interface itself to approaches oriented towards objects or to the tasks carried out by the user. *OO-Method* [212], for example, is a design paradigm oriented towards objects for the development of software systems based on development settings directed by models (*Model-Driven Development*). In *OO-Method*, the interaction elements are defined using a set of interaction patterns expressed in the pattern language *Just-UI* [189]. Another solution based on pattern modeling on UML [142, 203] is *WISDOM* [198], a software engineering method for the construction and maintenance of interactive software applications in the context of small and medium-sized enterprises (SME's). A third case constitutes *UsiXML* [164]. These examples in software engineering, show the advantages of using the pattern concept for the formal representation of presentation and interaction mechanisms, with some limitations. OO-Method, for example, is limited to the sphere of application of interfaces based on forms, whereas the application of WISDOM is recommended by its authors in *Small Software Developing Companies* (SSDs) to represent interfaces that show small sets of data in the form of reports, and not large volumes of data for decision making. Furthermore, all of them show a strong dependence between the formal representation of presentation and interaction and other models which illustrate other aspects of the system, such as function and persistence models (and requiring specific interaction modeling skills on the part of the specialist). This dependence limits the capacity of application of the patterns identified to other cases, such as their direct integration to the framework presented here or to any other solution designed to provide software assistance.

Finally, RIA (*Rich Internet Applications*) patterns are reusable solutions for common problems in the specification of presentation and interaction in interfaces [89]. RIA patterns present an approach which includes the advantages of the design based on patterns, as well as improving the reusability of the proposed solutions. However, their direct application as a solution for the representation of presentation and interaction in the proposed framework is not possible, due to the fact that they are traditionally applied in the definition and formal representation of collaborative contexts of application or social networks [34]. RIA patterns can be found, for example, in user profiles and in search integration (the handling of active, inactive or recommended searches). For this reason, it is necessary to define new RIA patterns within the framework of this research, which are specific for software assistance to the analysis of data and decision-making processes.

As all these works revealed, the specification of interaction based on patterns presents great advantages: an integral treatment of presentation and interaction mechanisms (thus avoiding ad hoc specifications), no excessive connection between the implementation and formal specification, allowing for the reuse of the solutions and avoiding the steep learning curve of specification languages.

Due to these advantages, the use of RIA patterns fits into our purpose of offering a solution for the formal representation of software presentation and interaction in applications focused on providing assistance to the analysis of data and decision making, and applying it specifically to our framework in order to assist to the archaeological generation of knowledge.

Some Challenges in Archaeological Data Analysis

This section will analyze some current challenges as far as the specification of software presentation and interaction for assistance to data analysis and decision making is concerned, paying special attention to interaction aspects. These challenges have been identified following the prior analysis of applications [179] used by archaeologists for data analysis and decision making based on that data, as well as via interviews with specialists in the discipline. The systems studied share characteristics in terms of data, processes and visualization, with the aim of assisting to the analysis of data in datasets. The challenges identified have been classified in two areas according to their origin: those related with the skills of the software analyst, domain specialist or user (note that in our case and through all the chapter, these terms refer to the archaeological specialists who are deciding how to visualize their data for analysis) and technological challenges.

As far as challenges related to the skills of the archaeological specialist are concerned, it is common that existing software tools for visualizing and representing in different ways archaeological data presents a steep learning curve for archaeologists. Our challenge as part of the framework is to facilitate a repository of easily understandable solutions for them with a context of application and one or several scenarios of use in which the advantages of using them can be illustrated.

This can be done via the election of patterns as a mechanism for presentation and interaction representation. Thus, the specialist, although he/she may not be an expert in interaction modeling, have at his/her disposal a repository of solutions which are fully adapted for software which provides assistance to data analysis and decision-making processes.

As far as the technological challenges are concerned, we have identified six factors in the definition of interaction for contexts regarding assistance to data analysis:

- CHALLENGE 1: The need to model presentation solutions for large volumes of data, thus solving spatial limitations on the screen and interface.
- CHALLENGE 2: The need for dynamic modeling and interaction solutions which allow for the dynamic visualization of data depending on the predominant (intentional) reasoning of the user at each moment.
- CHALLENGE 3: The need for modeling to deal with levels of importance: the same set of data can play different roles depending on the task being carried out by the user at each moment.
- CHALLENGE 4: The need for modeling of the use of different elements (colour, size, etc.) as interaction resources (solutions) in order to assist the user in data analysis tasks.
- CHALLENGE 5: The need for modeling of intrinsic characteristics of certain types of data, especially data of a geographic or temporal nature.
- CHALLENGE 6: The need for modeling schematic presentation solutions of data with sequential relationships between them, in such a way as to visually maintain the sequence.

The challenge definition could be applied to other data domains, although the objectives of this framework is focused in archaeology. Thus, we require a solution based on design patterns with a sufficient degree of abstraction in order to deal with the challenges independently of application, though always contextualizing the solution within the objectives of providing software assistance to data analysis and decision-making processes. The following section illustrates the design patterns for dealing the challenges in an abstract way, and subsequently explains in detail how applying them for archaeological data interaction representation.

Design Patterns for Archaeological Data Analysis

At this point, it is necessary to highlight that, although there have been attempts within the research community to reach agreements in the notation and definition of patterns, there is currently no standard which allows us to define in a single way a solution of these characteristics. This situation leads to the lack of a single way to define the proposal of patterns we shall put forward below which constitutes our solution for the representation of presentation and interaction for software assistance (to be integrated later via a specific implementation into our framework for archaeological data sets). RIA model serves us here to describe the patterns, without prejudicing other models of pattern definition which could also be used. The most commonly used structure in the definition of RIA patterns consists of four parts [194]: a title, an identified problem to be resolved, an application context and a solution. Furthermore, we have included a fifth element in the specifications of our solution which describes scenarios of use for each one of the patterns, in an attempt to gain a better overall understanding of the solution by the readers.

In addition, existing studies in this area have identified the need to work on this type of proposal with different levels of abstraction in order to define a pattern solution [188]. The classification of the solutions from the most abstract to the most specific level allows behaviors to be encapsulated throughout the different levels, with the most simple patterns being able to be reused in order to specify more complex patterns [188]. Following this approach, three levels of patterns have been defined:

- LEVEL 1 Data-Analysis Assistance Unit: This consists of just one interaction unit which acts as a containing mechanism. Thus, the Data-Analysis Assistance Unit encapsulates the available units of interaction in order to assist specialists in decision making based on data. This pattern is an abstract representation of a navigational menu by way of interfaces. It should be noted that due to the fact that we only have one element in level 1 of the pattern hierarchy, this level will not be described in greater detail at a later point.
- LEVEL 2 Interaction Units: An Interaction Unit (hereinafter referred to as IU) is an abstract representation of a complete interface which will be used by specialists in order to carry out data analysis and decision-making tasks. Each IU can be seen as a set of presentation methods and simple behaviors identified in the third level for the support of a certain cognitive process or data analysis task.
- LEVEL 3 Individual Patterns: Each individual pattern identifies presentation representations and interaction behaviors which can be used within different Interaction Units. An individual pattern is an abstract representation of an interface widget with a predefined specific behavior.

All of the patterns which make up the hierarchical suite which we propose as a solution can be seen in Fig. 9.1. Below, we shall describe each pattern following the base structure in order to define the previously mentioned RIA patterns.

LEVEL 1: Data-Analysis Interaction Unit

- Problem: A need to represent as a whole the presentation and interaction mechanisms aimed specifically at assisting domain specialists in data analysis and decision making.

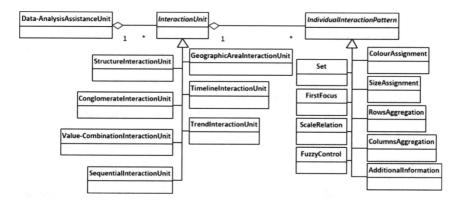

Fig. 9.1 Metamodel expressed in UML representing the proposed interaction and presentation patterns in the framework proposed

- Context of application: Situations identified as requests by specialists in the pursuit of bringing together software assistance solutions for data analysis and decision making.
- Solution: Defining a first containing level which permits the grouping of the chosen solutions in each case to represent the software presentation and interaction, allowing for its reuse in other systems, architectures, configurations and designs of assistance applications. This level does not respond to any specific CHALLENGE but it attempts to provide the proposed solution with a grouping mechanism which allows for the reuse of solutions, as was previously identified in this chapter, as an objective of the solution to be proposed.
- Scenario of use: Having at his/her disposal a mechanism for grouping solutions, the specialist can reuse a set of solutions which have already been selected for similar applications or purposes, thus encapsulating them.

LEVEL 2: Structure IU

- Problem: The specialists being assisted need a general view of the structure of data used to define the information contained in a certain dataset to be visualized, emphasizing the structure of the information and not the information itself.
- Context of application: Situations identified as requests by the specialists in the pursuit of improving understanding and self-awareness regarding the underlying structure to the information which is to be analyzed, according to an OO (Object-Oriented) paradigm (): classes, attributes and associations.
- Solution: To organize the information in an interaction unit following OO (Object-Oriented) criteria (classes, attributes and associations) but also hiding technical aspects from the archaeological specialists (types of data and technical

specifications of the database). The interaction unit should be capable of representing the structure of the information, as well as maintaining a visual central theme with the specific instances, being able to visualize how that structure is reflected in a specific instance. This double view offers them a clear view of the structure of the dataset, allowing him/her to play with the abstraction and to visualize the possible implications on the level of instances which any structural modification will have in the information. This solution is directly related with CHALLENGE 1.

- Scenario of use: An analysis based on the structure of information is a common practice in applications aimed at providing software assistance to data analysis. For example, the specialist can see the attributes of different classes in order to decide if it would be interesting to compare these classes, their common structure or to establish what criteria are most appropriate for a later classification on the level of instances.

LEVEL 2: Value-Combination IU

- Problem: The specialists being assisted need interaction and presentation mechanisms for searches and the evaluation of instances depending on the values of their attributes.
- Context of application: Situations identified as requests by the specialists in the pursuit of offering interaction solutions which enable us to know the values of the different attributes, classifying the information according to those values.
- Solution: To organize the information contained in a dataset, allowing for the election of a main class (which acts as an objective class for the analysis) and to classify the instances of that class depending on the values of its attributes. This solution is directly related with CHALLENGES 1, 2 and 4.
- Scenario of use: The analysis based on values of attributes allows the specialist to reason about characteristics of a statistical nature present in the dataset, such as inferring averages, relevant percentages, etc.

LEVEL 2: Conglomerate IU

- Problem: The specialists being assisted need interaction and presentation mechanisms which allow the data contained in the dataset to be classified in a dynamic and simple way.
- Context of application: Situations identified as requests by the specialists in the pursuit of offering greater dynamism in the rapid classification of the information contained in the dataset by a dynamic criterion.

- Solution: To organize the information contained in a dataset, permitting the selection of the classifying criterion and emphasizing the speed of reconfiguration of the interface. This solution is directly related with CHALLENGES 2 and 4.
- Scenario of use: An analysis based on rapid classifications of a large volume of data by just one criterion helps the specialist to understand which main entities exist in the dataset being analyzed, as well as deviations in the data. For example, he/she can detect atypical groups with respect to a certain criterion or instances which do not belong to any group (extreme values, etc.).

LEVEL 2: Trend IU

- Problem: The specialists being assisted need interaction and presentation mechanisms which allow tendencies in the data in quantitative terms to be observed, both within one dataset (for example, the number of entities or instances which form part of one group or another) and in other datasets which have the same structure of information (for example, the number of entities or instances present in two datasets from different sources or two versions of the same dataset taken at different moments).
- Context of application: Situations identified as requests by the specialists in the pursuit of offering interaction solutions for the detection of quantitative tendencies in medium and large volumes of data.
- Solution: To organize the information contained in a dataset in quantitative terms, allowing for the selection of a grouping criterion. The interface should be able to structurally recognize the information being presented in order to be able to visualize two datasets at the same time in quantitative terms, providing that the structure of the information visualized in both datasets coincides. This solution is directly related with CHALLENGES 3 and 4.
- Scenario of use: An analysis based on the detection of quantitative tendencies present in a large volume of data helps the specialist to understand what factors intervene in the belonging of objects of the dataset to a certain group, to compare data with the same structure coming from different versions and to infer possible future behaviors in the data. For example, a specialist can detect the importance of a certain group of data (by its number of elements) due to the fact that, in successive versions of the same dataset, this group increases its number of objects or he/she can compare two datasets from different sources in order to know whether their groups behave in the same way or whether they present different tendencies (one rising and the other falling in number of elements, etc.).

LEVEL 2: Timeline IU

- Problem: The specialists being assisted need interaction and presentation mechanisms which allow the temporal aspects of the data to be analyzed, especially attributes of a temporal nature or classes with values in attributes which change over the course of time.
- Context of application: Situations identified as requests by the specialists in the pursuit of permitting the analysis of changes over the course of time in values of attributes or instances of classes with a temporal component.
- Solution: To organize the information in order to select, visualize and interact with values of variable attributes over time in an interface which is particularly aimed at visual reasoning. This solution is directly related with CHALLENGE 5.
- Scenario of use: The analysis of temporal data allows the specialist to infer temporal dependencies in the data. For example, he/she can see how the values of two attributes change over the course of time in order to analyze possible relations between both changes.

LEVEL 2: Geographic Area IU

- Problem: The specialists being assisted need interaction and presentation mechanisms which allow geographical aspects of the data to be analyzed, especially the attributes of a geographical nature or classes with eminently geographic semantics (locations, places, etc.).
- Context of application: Situations identified as requests by the specialists in the pursuit of permitting the geographic analysis of values of attributes or instances of classes with a geographic component.
- Solution: To organize the information in order to select, visualize and interact with the information contained in an interface which is especially oriented towards geographic reasoning. This solution is directly related with CHALLENGE 5.
- Scenario of use: A geographic analysis allows the specialist to situate his/her data geographically. For example, he/she can see the data according to its geographic location of origin, being aware of the geographic area which it covers, thus allowing for the analysis of this coverage and/or the detection of possible implications.

LEVEL 2: Sequential IU

- Problem: The specialists being assisted need interaction and presentation mechanisms which allow data connected with sequential relations to be analyzed, be it a sequence of data of the same nature or any information which must be visualized "by levels" in order to be understood.
- Context of application: Situations identified as requests by the specialists in the pursuit of offering interaction solutions for sequential or organized information visualization in levels or layers.
- Solution: To organize the information in order to select, visualize and interact with the information contained in an interface which is especially oriented towards sequential reasoning "by levels". This solution is directly related with CHALLENGES 3, 4 and 6.
- Scenario of use: A sequential analysis allows the specialist to situate his/her data in levels. For example, he/she can see the data belonging to a sequence along with its relations, thus allowing him/her to analyze these relations and/or detect possible implications.

LEVEL 3: Row Aggregation Pattern

- Problem: The specialists need a general view of the information contained in a dataset organized into interfaces with a landscape orientation.
- Context of application: The specialist requests a general view of the information, for example, according to values, or their intervals, of an attribute or of classes belonging to different instances, but always maintaining the landscape orientation.
- Solution: To organize the information in rows, creating visual divisions in the interface which allow the rows to be identified clearly and intuitively, without overloading the interface. This allows an analysis to be carried out by the specialist in the direction of reading of the interface (horizontally) but without affecting the direction of the reading. In other words, the interface can be read from right to left and vice versa, maintaining the distinction between the categories and intervals of values represented in the rows. This solution is directed related with CHALLENGE 1.
- Scenario of use: A horizontal visual analysis, maintaining the organization in rows, allows the presentation of a large amount of data to be adapted in an organized and convenient manner for human visual analysis.

LEVEL 3: Column Aggregation Pattern

- Problem: The specialists need a general view of the information contained in a dataset organized into interfaces with a vertical orientation.
- Context of application: The specialist requests the visualization of a general view of the information, for example, according to values, or their intervals, of an attribute or of classes belonging to different instances, but always maintaining a vertical orientation.
- Solution: To organize the information into columns, creating visual divisions in the interface which allows the columns to be identified clearly and intuitively, without overloading the interface. This allows an analysis to be carried out vertically, maintaining the distinction between the categories or intervals of values represented in the columns. This solution is directly related with CHALLENGE 1.
- Scenario of use: A vertical visual analysis, maintaining the organization in columns allows the presentation of a large amount of data to be adapted in an organized and convenient manner for human visual analysis.

LEVEL 3: Set Pattern

- Problem: The specialists need to visualize a dataset grouped according to a criterion, treating the resulting groups as elements of the interface with an entity of their own.
- Context of application: The specialist requests the visualization of aggregate information, generally according to a criterion, requiring the interaction and treatment of the resulting groups.
- Solution: To organize the information creating an element of interface, generally a recognizable shape (a sphere, an ellipse, etc.) for each group resulting from the aggregation. This solution is directly related with CHALLENGE 2.
- Scenario of use: An analysis based on categories requires the choice of an interface element as a graphic notation representing the groups of data of the different categories created.

LEVEL 3: Additional Information Pattern

- Problem: The specialists need to obtain additional information about data or datasets which are present in the interface.
- Context of application: The specialist requests the visualization of contextual information of some specific data or a dataset of the interface in which it is found, albeit without any changes in the interface.

- Solution: To organize the information by showing specific information contextualized to the element which the specialist wishes to know more deeply. This information is shown with an additional element, which does not intrude in the interface and which is activated in response to an action carried out by the specialist (typically a *click* or a *mouse hover*). This solution is directly related with CHALLENGE 3.
- Scenario of use: An analysis of data in its context with small enquiries for contextual information is common in data analysis processes and allows the specialist not to lose the overall vision of the data being analyzed, allowing for knowledge to be gained via interaction with the interface of more specific information of the elements of which it consists. This additional information, which is offered quickly and unobtrusively, allows him/her, for example, to gain a deeper knowledge of data which is out of range or which has an abnormal behavior and to infer possible causes.

LEVEL 3: First Focus Pattern

- Problem: The specialists need to filter visual noise in a complex interface showing a large amount of data.
- Context of application: The specialist requests the visualization of several focuses or parts in an interface with a high density of presented data, thus avoiding the visual noise which can be caused by the other elements of the interface.
- Solution: To organize the information into two levels, so that through the use of shading or other highlighting or visual deactivation mechanisms, the specialist can focus on one part of the presented information. This solution is directly related with CHALLENGE 3.
- Scenario of use: Interfaces with a high density of data generally require focused visualization in one or more places at the same time in order to establish comparisons or to highlight parts or aspects of the information. This pattern offers a solution to cover these analysis needs without structural changes in the main interface.

LEVEL 3: Colour Assignment Pattern

- Problem: The specialists need mechanisms in order to give different semantic content to each element of the interface.

- Context of application: The specialist requests the identification of different elements of the interface with semantic coherence regarding the data which it represents.
- Solution: To organize the information offering the specialist a mechanism to choose a colour as the differentiating and representative element of elements of the interface and, therefore, of aspects and datasets which are shown within it. If the specialist groups information according to a criterion, for example, he/she can use colour in order to associate each group to a semantic value of that criterion. This solution is directly related with CHALLENGE 4.
- Scenario of use: The selection of colour is an effective mechanism to provide meaning to the interface in an unobtrusive way, also allowing a visual logic to be maintained throughout various interfaces of data analysis.

LEVEL 3: Size Assignment Pattern

- Problem: The specialists need mechanisms to give semantic content to the relations between each element of the interface and their size within the interface.
- Context of application: The specialist requests the association of the size of the different elements of the interface with semantic coherence regarding the data which they represent.
- Solution: To organize the information offering a mechanism to choose what meaning the size of the interface's elements will have. If the specialist groups information according to a criterion, for example, he/she will be able to use size to associate each element of the interface (which represents each resulting group) with its size according to the number of elements which make up each group. This solution is directly related with CHALLENGE 4.
- Scenario of use: Using the size of elements of the interface is an effective mechanism to provide meaning in an unobtrusive manner, also allowing a visual logic to be maintained throughout various interfaces for data analysis.

LEVEL 3: Scale Relation Pattern

- Problem: The specialists need to visualize information, generally values of attributes, whose nature is temporal or varies over the course of time.

- Context of application: The specialist requests the visualization of data with a strong temporal component, such as phases or events relating to instances in a temporal context. Occasionally, he/she requests this visualization separately, in such a way that various attributes or data of a temporal nature can be represented in the same interface.
- Solution: To organize the information offering a mechanism which relates various temporal datasets in the same interface. This solution is directly related with CHALLENGE 5.
- Scenario of use: The solutions for visualizing several temporal datasets at the same time generally consist of visualizing them separately and combining them in the interface. This pattern allows the implicit relations between several sets of temporal information to be taken into account and to visualize them in different timelines, whilst maintaining the visual relation between them.

LEVEL 3: Fuzzy Control Pattern

- Problem: The specialists need to visualize information, generally values of attributes, the precision of which varies greatly from one to another, showing intervals with a great degree of diffusion in their values.
- Context of application: The specialist requests the visualization of data whose values present different degrees of diffusion (wide intervals compared with very specific values) and a certain degree of imprecision (diffuse values on their limits).
- Solution: To organize the information offering a mechanism which, given a value of an attribute, allows its degree of diffusion to be visualized. It is generally implemented by making the element of the interface which represents the value vibrate. If the element covers more space when vibrating, the extent of the interval in values is greater and vice versa. In addition, if the element presents a lower frequency of vibration, it is more diffuse (understanding by 'diffuse', less precise in its definition), whereas if it vibrates with greater frequency, the diffusion is less. This solution is directly related with CHALLENGE 5.
- Scenario of use: Diffusion in the values is a common characteristic in certain types of data to be analyzed, which should be represented in order to assist to data analysis and decision making. This pattern allows the extent of the values of an attribute to be taken into account, along with the degree of diffusion which that value presents.

Table 9.1 sums up the interaction units defined and the challenges addressed:

Table 9.1 Matrix of challenges versus units of interaction and individual patterns which each challenge deals with

	CHALLENGE 1	CHALLENGE 2	CHALLENGE 3	CHALLENGE 4	CHALLENGE 5	CHALLENGE 6
2. Structure IU	•					
2. Value-combination IU	•	•		•		
2. Conglomerate IU		•		•		
2. Trend IU			•	•		
2. Timeline IU					•	
2. Geographic area IU					•	
2. Sequential IU			•	•		•
3. Row aggregation	•					
3. Column aggregation	•					
3. Set		•				
3. Additional information			•			
3. First focus			•			
3. Colour assignment				•		
3. Size assignment				•		
3. Scale relation						
3. Fuzzy control					•	

Applying Design Patterns in Archaeology

As can be observed, the set of challenges identified and solutions proposed in the form of the RIA patterns described above is applied to the set of software assistance applications for data analysis. Due to the fact that our studies [103, 105, 174] have been carried out in Archaeology, we have refined them as specific problems in the discipline, using the same enumeration as the challenges defined abstractly and independently of the field. In this section, we shall present the problems contextualized in terms of archaeological data analysis.

PROBLEM 1: Archaeologists and related specialists often work in teams on one particular dataset [104, 179]. Therefore, they require a general view of the structure of the information with which they are working, which will be a determining factor in order to clarify its content, helping to decide the strategy for analysis and research to be followed regarding the archaeological data. This general view should show both the structure of the information and its relation with the data itself, which supposes the visualization of a large amount of data at the same time. One example would be the need for a general view of the structure of the information of a dataset from an archaeological project or from a set of anthropological interviews, etc. This general view should assist the archaeologist in understanding the structure and should also give him/her information on specific instances (a specific interview, an archaeological finding in that particular project, etc.).

PROBLEM 2: The process of categorization or grouping is extremely common in Archaeology, forming part of the working methodology in the different fields of which it consists. Some good examples of this include the construction of thesauri, agreements regarding terminology and the typology of objects and evidence. It is necessary, therefore, to have a presentation and interaction mechanism which enables the simple and agile grouping of the data present in the dataset according to a criterion, which the archaeologist may vary.

PROBLEM 3: Archaeologists and related specialists often work with a dataset in which they need to deal with different levels of importance of the information in question. Some good examples of this include datasets containing information about archaeological evidence, historical events, literary references, etc. In all of these cases, the need arises to obtain additional information about the evidence, event or reference or a subset of them, whilst maintaining the general view of all of them on the screen.

PROBLEM 4: Within the process of analysis in archaeology, it is common to find the presence of several attributes which cut across the dataset which the archaeologists are analyzing to detect similarities or differences between the data. Good examples of this include morphology or decoration in archaeological evidence, the cultural assignation of historical or literary instances, the materials of which objects are made, etc. These similarities and differences should be made clear by way of the use of interface mechanisms which allow them to be visible.

PROBLEM 5: In archaeology, temporal and geographic components of data are the main axes for analysis in order to be able to postulate scientific hypotheses and to make decisions based on the data which has been observed [179]. In this context, the interfaces which present this data should reflect each one of the facets in a way which is adapted to the type of analysis which archaeologists and related specialists commonly carry out, each with their own problems, such as the dispersion of temporal data or the support for spatial reasoning. A good example of this could be the needs for the temporal visualization of events or phases with differing degrees of dispersion in historical and/or archaeological contexts.

PROBLEM 6: Within the process of the categorization of information reflected in the problems detected above, some datasets exist in archaeology with a strong sequential (or level-based) component. The sequential structure of this type of information could go unnoticed without an interface which explicitly includes this type of data and the relations between its instances. Some good examples of this include stratigraphic studies, spatial analysis of buildings and formation, geological strata, etc. Indeed, sequential schematic visualization could arise in any area of Cultural Heritage.

In conclusion, the abstract challenges identified arise as problems when we want to represent presentation and interaction mechanisms for software assistance to the analysis of archaeological data. We shall now go on to describe each interaction unit which has been implemented and the individual patterns chosen for assistance when analyzing archaeological data sets.

Structure IU

The interaction unit Structure IU responds to Problem A identified above in archaeological data analysis (and related with the general challenge 1). It makes reference to the need to provide the archaeologists with a general view of the structure of the information, whilst maintaining a connection with the specific data. In order to achieve this, an interface has been designed which allows the whole underlying structure in a dataset to be visualized, thus avoiding methodological bias in its visualization (diagrams corresponding to different disciplines or branches of information sciences, such as Entity-Relationship diagrams [62] or class diagrams [142] and the notations already used for them). This interface presents two ways of working: in Classes and Instances modes.

In Classes mode, the interface, given a dataset chosen by the archaeologist (the user of the interface at that moment) and a main class, shows the structural relationships of the selected class with the rest of the classes of the model. This allows to "position" him/herself in one part of the structure of the dataset's information and to navigate through it. If the archaeologist wishes, he/she can also access the defined attributes for that class, thus accessing not only a general view of the dataset's structure but also knowing what information he/she can store in it.

In Instances mode, the interface maintains the design of the previous mode but permits to select a specific instance of the dataset and to navigate through its relationships whilst maintaining information about the structure of the information.

If, whilst navigating, he/she finds another instance, this will be accessible. If, on the other hand, no object is instanced, information regarding the corresponding class will be maintained but the object will not be accessible, thus permitting the archaeologist to maintain, at all times, an overall vision of what classes have associated instances. From this interface it is possible to access the values of each instance's attributes.

Internal Components

Four individual patterns have been selected from level 3 in order to compose our Structure IU proposal:

– Colour assignment: This pattern has been selected to use the colour mechanism in the differentiation of the nature of the classes or instances within the structure of the information. More specifically, Fig. 9.2 shows how the colour will differentiate the classes of the reference model which is used throughout this research (CHARM [99], explained in detail in Chap. 7), its extended classes, the elements of the structure of the information with a temporal component and those which have a subjective component.

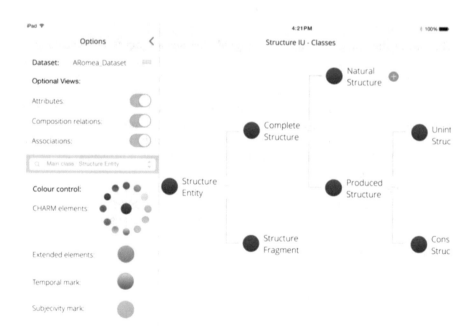

Fig. 9.2 Structure IU-classes: the configuration options can be seen on the left

- Additional Information: This pattern has been selected to show additional information about the relations between classes and instances in the Structure IU interface, basically providing information about what type of relationship is present and whether it has a temporal and/or subjective aspect.
- First Focus: This pattern has been selected to offer visual clarity in the presentation of additional information about the relations between classes and instances in the Structure IU interface. When the archaeologist selects a relation, the interface changes its focus to that relation, fading the background in order to highlight the element.
- Row Aggregation: This pattern has been selected in the design of the interface in order to show the structure of the information. Due to the fact that complex and/or profound structure can be presented, requiring a lot of space on the screen, the decision was taken to promote horizontal visualization, playing with the horizontal scroll in order to access the full hierarchy. This allows the classes of the same level to be maintained together on the screen. Furthermore, in the application of the Row Aggregation pattern, the visualization of the limits of each row has been discarded, mainly due to reasons of clarity, minimalism in the design and minimization of visual noise.

Aspects of Use and Interaction

The aim of this interaction unit is to provide the archaeologists and related specialists with a flexible presentation and interaction mechanism, based on the specialist's reasoning about the structure of the information. Figures 9.2, 9.3, 9.4 and 9.5 show a visual prototype. The nominal sequence of steps will consist of the archaeologist selecting a dataset and a main class within it. Then, he/she will have to decide what type of structural information is of interest: attributes, composition and/or association relations. Finally, he/she will be able to decide the colours he/she wants to differentiate the different aspects of the information explained above by using the colour control.

Once the interface has been configured, he/she can navigate through the underlying structure of the information in the dataset. If an association is selected, as can be seen in Fig. 9.4, the First Focus and Additional Information patterns will help to know the nature of the association. If a class is selected, access will be provided to its defined attributes. The class hierarchies can be expanded and contracted in order to avoid visual noise in very large and/or deep structures of information and, at any time, a different class can be selected as the main class and the archaeologist can resituate him/herself in the structure of the information.

This nominal sequence of steps is identical for instances. Generally, the natural option is to access Structure IU—Classes in order to then select in Structure IU—Instances a specific instance of interest for the archaeologist, once he/she knows the structure of the information in that part of the model. Once selected, he/she will

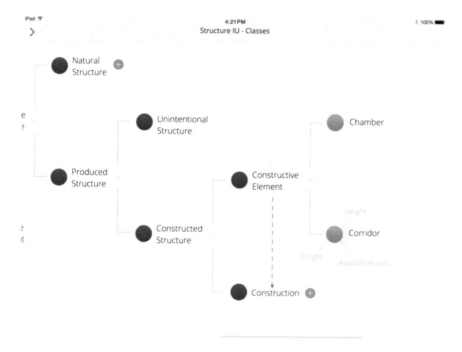

Fig. 9.3 Structure IU-classes: defined attributes for a selected class

navigate in a similar way to the Structure IU—Classes interface but, this time, accessing the specific values for that instance of the corresponding attributes. The interface in Structure IU—Instances mode can be seen in Fig. 9.5.

Value-Combination IU

The interaction unit Value-Combination IU deals with problems A, B and mainly D (and related with the general challenges 1, 2 and 4) identified previously for the archaeological data analysis. In this implementation, we have focused on problem D, which makes reference to the need to detect similarities or differences between the present data, which are shown in values of attributes. In order to do this, an interface has been designed which allows one main class in a chosen dataset to be selected and the groups of instances which form to be visualized according to several values of attributes (up to three combined attributes at one time). This enables the archaeologists to search for similarities and differences in the data as it groups the data which have the same values into certain attributes.

Fig. 9.4 Structure IU-classes: an example of the behavior of the additional information and first focus patterns

Internal Components

Seven individual patterns from level 3 have been selected to compose our Value-Combination IU proposal.

- Size assignment: This pattern has been selected in order to use the size mechanism, associating the size of the elements of the interface to the perception of the size of the groups which the archaeologist forms in the interface by combining values of attributes.
- Colour assignment: This pattern has been selected in order to use the colour mechanism in the differentiation of a specific attribute within the combination of values of the interface. For example, Fig. 9.6 shows how the colour will differentiate the data present in the dataset by way of instances of Object Fragment according to the material of each fragment of object. This attribute has already been selected to be combined in the rows but it can be highlighted thanks to the colour. It would be possible to choose another attribute which has not been selected previously in order to illustrate it with the colour mechanism.
- Additional Information: This pattern has been selected in order to show additional information about the groups formed by the archaeologist in the interface by combining values of attributes.

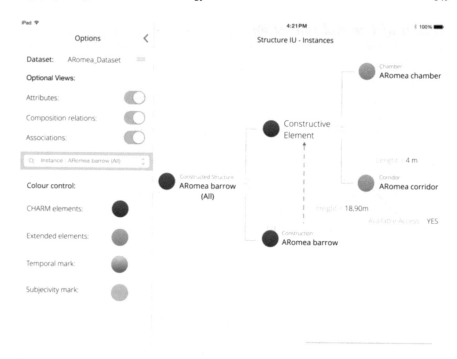

Fig. 9.5 Structure IU-instances: the configuration options can be seen on the left

- First Focus: This pattern has been selected in order to offer visual clarity in the presentation of additional information about the groups formed by the archaeologist in the interface when combining values of attributes. He/she selects a group and the interface changes the focus of that element, fading the background in order to highlight the element.
- Row Aggregation: This pattern has been selected in the design of the interface in order to visually organize the categories or intervals of one of the attributes to be combined in the Value-Combination IU interface.
- Column Aggregation: This pattern has been selected in the design of the interface in order to visually organize the values (or their intervals) of an attribute or the classes to which the different instances to be combined belong in the Value-Combination IU interface.
- Set: This pattern has been selected in order to represent the groups formed by the archaeologist when combining values of attributes as interface elements with entities of their own. In this case, a bubble shape has been chosen for the implementation of the Set pattern.

Aspects of Use and Interaction

The aim of this interaction unit is to provide the archaeologists and related professionals with a flexible presentation and interaction mechanism, based on the specialist combining values of attributes in order to find similarities and differences. Figures 9.6 and 9.7 show a visual prototype. The nominal sequence of steps will consist of the archaeologist selecting a dataset and a main class within it. Then, he/she will have to decide what attributes are of interest and, by using the colour control, select the colours he/she wishes to use in order to differentiate another attribute or the values which the data presents according to a previously chosen attribute. Finally, he/she will be able to decide what semantics will be implicit in the size of the interface's elements (a fixed size, according to the number of instances of information belonging to each group, etc.).

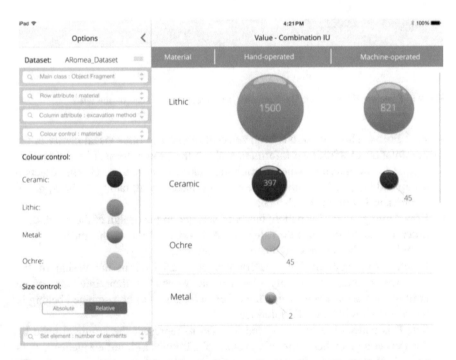

Fig. 9.6 Value-combination IU: the configuration options can be seen on the left

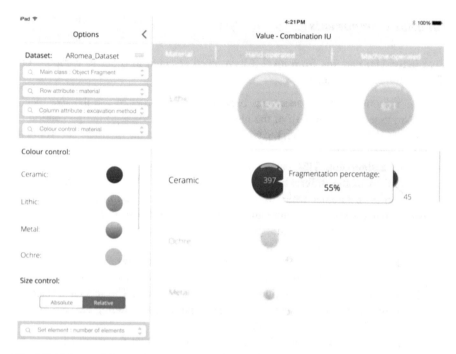

Fig. 9.7 Value-combination IU: an example of the behavior of the additional information and first focus patterns

Conglomerate IU

The interaction unit Conglomerate IU deals with problems B and D (and related with the general challenges 2 and 4), previously identified for archaeological data analysis. In this implementation, we have focused on problem B, which makes reference to the need for a presentation and interaction mechanism which allows for the simple and agile grouping of the data present in the dataset according to one criterion, which the archaeologist can vary. In order to do this, an interface has been designed which allows one main class in a chosen dataset to be selected and the groups of instances which are formed according to a selected criterion to be visualized. In this way, the dynamism in the action of grouping is maintained, due to the fact that the archaeologist can change the grouping criterion with a simple action (typically a click).

Internal Components

Five individual patterns from level 3 have been selected in order to compose our Conglomerate IU proposal:

– Size assignment: This pattern has been selected in order to use the size mechanism, associating the size of the elements of the interface to the perception of the size of the groups which the archaeologist forms in the interface according to the selected criterion.
– Colour assignment: This pattern has been selected in order to use the colour mechanism in the differentiation of the groups formed according to the main criterion or to another which is applicable to the same main class.
– Additional Information: This pattern has been selected in order to show additional information about the groups which the archaeologist forms in the interface according to the selected criterion.
– First Focus: This pattern has been selected in order to offer visual clarity in the presentation of additional information about the groups formed by the archaeologist in the interface according to the selected criterion. The specialist selects a group and the interface changes the focus of that element, fading the background in order to highlight the element.
– Set: This pattern has been selected in order to represent the groups formed by the archaeologist according to the selected criterion as interface elements with entities of their own. In this case, a bubble shape has been chosen for the implementation of the Set pattern.

Aspects of Use and Interaction

The aim of this interaction unit is to provide the archaeologists and related professionals with a flexible presentation and interaction mechanism, based on the formation of groups of instances of a dataset in an agile and rapid manner. Figures 9.8 and 9.9 show a visual prototype. The nominal sequence of steps will consist of the archaeologist selecting a dataset and a main class within it. Then, he/she will have to decide which grouping criterion is of interest and, by using the colour control, select the colours he/she wishes to use in order to differentiate the groups formed according to the main criterion of another attribute of the main class. Finally, he/she will be able to decide what semantics will be implicit in the size of the interface's elements (a fixed size, according to the number of instances of information belonging to each group, etc.).

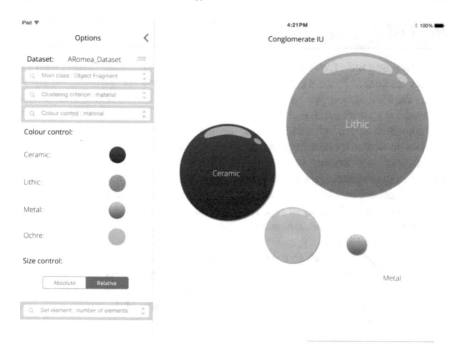

Fig. 9.8 Conglomerate IU: the configuration options can be seen on the left

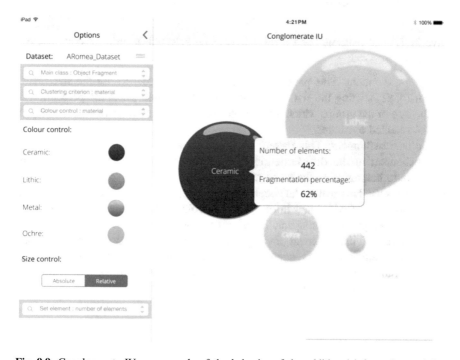

Fig. 9.9 Conglomerate IU: an example of the behavior of the additional information and first focus patterns

Trend IU

The interaction unit Trend IU deals with problems C and D (and related with the general challenges 3 and 4), previously identified for archaeological data analysis. In this implementation, we have focused on Problem C, which makes reference to the need for a presentation and interaction mechanism which allows different levels of importance in the information to be dealt with. Often, this importance is not possible to detect by studying the structure of the information or the groups of which it consists. Rather, we need a mechanism which allows us to observe tendencies in the datasets and subsets in order to then determine what roles each part of the information plays in a given dataset. Therefore, an interface has been designed which allows a main class in a chosen dataset to be selected and, optionally, a second dataset which maintains the structure corresponding to that main class and its relations. This secondary dataset is, typically, a previous or later version of the first one or a dataset which contains data belonging to another source which shares the same information structure as the main dataset. Once a dataset and a main class have been selected, the archaeologist can select the attributes about which he/she wishes to observe tendencies and visualize the groups of instances and their importance in the dataset.

Internal Components

Five individual patterns from level 3 have been selected in order to compose our Trend IU proposal:

- Size assignment: This pattern has been selected in order to use the size mechanism, associating the size of the elements of the interface to the perception of the size of the groups which the archaeologist forms in the interface according to the selected attributes.
- Colour assignment: This pattern has been selected in order to use the colour mechanism in the differentiation of the groups formed according to a third attribute belonging to the same main class.
- First Focus: This pattern has been selected in order to offer visual clarity in the presentation of additional information regarding the tendency in the data of a specific subgroup. The specialist selects an element and the interface changes the focus of that element, fading the background in order to highlight the specific element and its tendencies (lines which mark whether the data with that value of attribute have grown or decreased in number, along with the evolution if we are comparing with another, secondary, dataset).

- Column Aggregation: This pattern has been selected in the design of the interface in order to visually organize the data and its tendencies belonging to two different datasets (main and secondary).
- Set: This pattern has been selected in order to represent the groups formed by the archaeologist according to the attributes selected as interface elements with entities of their own. In this case, a bubble shape has been chosen for the implementation of the Set pattern. If it is not possible to view the bubble on the screen, it will vary, becoming an element in the shape of a drop in order to indicate that the position of the group corresponds to a value outside of the screen (by scrolling and accessing the data it will revert to the bubble shape).

Aspects of Use and Interaction

The aim of this interaction unit is to provide the archaeologists and related professionals with a flexible presentation and interaction mechanism, based on the detection of tendencies in the data and on the comparison of two versions of the same dataset or of datasets which share the same structure as far as the level of importance of the data is concerned. Figures 9.10, 9.11 and 9.12 show a visual

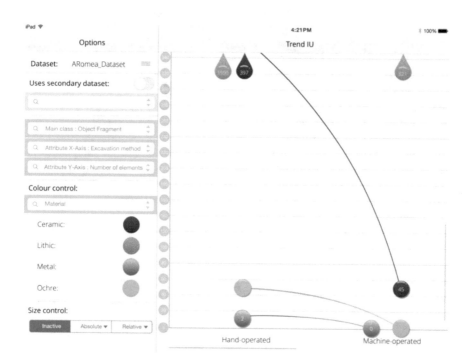

Fig. 9.10 Trend IU: the configuration options can be seen on the left

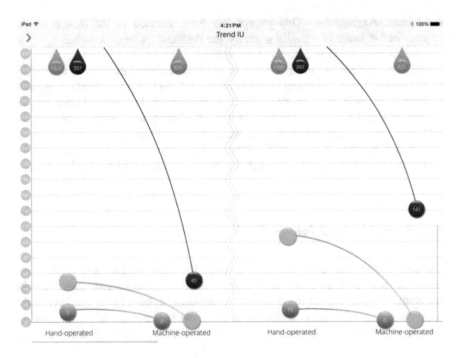

Fig. 9.11 Trend IU: an example of behavior for two given datasets

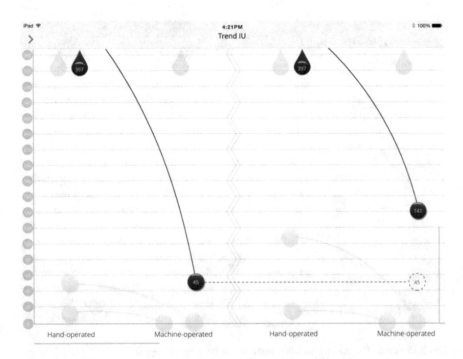

Fig. 9.12 Trend IU: an example of the behavior of the first focus pattern to assist in the detection of tendencies in two given datasets

prototype. The nominal sequence of steps will consist of the archaeologist selecting a dataset and a main class within it. If his/her aim is to compare two datasets, he/she can select the secondary dataset. Then, he/she will have to decide which attributes are to be used as information aggregates in order to observe the tendencies and, by using the colour control, select the colours he/she wishes to use in order to differentiate the groups formed according to the attributes chosen or to another attribute belonging to the main class. Finally, he/she will be able to decide what semantics will be implicit in the size of the interface's elements: a fixed size, according to the number of instances of information belonging to each group, etc. In this case, in Trend IU the decision has been made to enable the size control to be deactivated in an attempt to minimize visual noise in the interface.

Timeline IU

The interaction unit Timeline IU deals with problem E (and related with the general challenge 5), previously identified for archaeological data analysis, which makes reference to the need for a presentation and interaction mechanism which allows the temporal facet of the data to be reflected in a way which is adapted to that carried out in archaeology, with its own problems such as the dispersion of temporal data. In order to do this, an interface has been designed which allows a main class in a chosen dataset to be selected, along with one of its specific instances. Optionally, it is possible to select a second class and an instance belonging to it. From these classes and instances, the attributes defined as Time data type or temporal associations [129] according to the structure of the dataset's information will be uploaded into the interface. The archaeologist will be able to add the attributes which he/she wishes to visualize in the Timeline IU interface.

Internal Components

Four individual patterns from level 3 have been selected in order to compose our Timeline IU proposal:

- Colour assignment: This pattern has been selected in order to use the colour mechanism in the differentiation of the temporal attributes which are visualized in Timeline IU.
- Additional Information: This pattern has been selected in order to show additional information about each interface element present in Timeline IU.
- Scale Relation: This pattern has been selected in order to visually reflect the correspondence between the visual timeline of reference and the set of values of

a temporal attribute. As many temporal attributes from the same main class or from the selected secondary class can be added as desired. The Scale Relation pattern will offer a timeline for each attribute, maintaining the vertical visual connection between the corresponding temporal periods in order to favor comparison and temporal reasoning.

– Fuzzy control: This pattern has been selected in order to represent how diffuse the time intervals are in which each value of each attribute is represented. The interface element which represents each value will vibrate horizontally more or less strongly depending on the degree of diffusion of the value: for more specific values, more vibration, for more diffuse values, less vibration. The strength of the vibration determines the extent of the temporal interval in which we are situated.

Aspects of Use and Interaction

The aim of this interaction unit is to provide the archaeologists and related professionals with a flexible presentation and interaction mechanism, based on the temporal aspects of the data. Figures 9.13, 9.14 and 9.15 show a visual prototype.

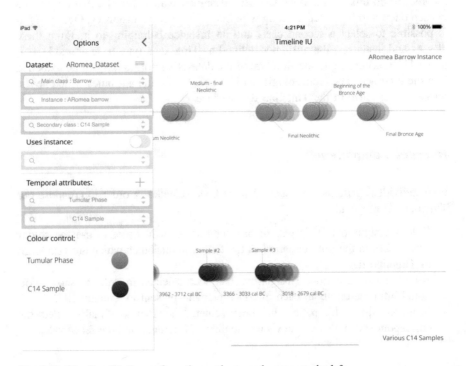

Fig. 9.13 Timeline IU: the configuration options can be seen on the left

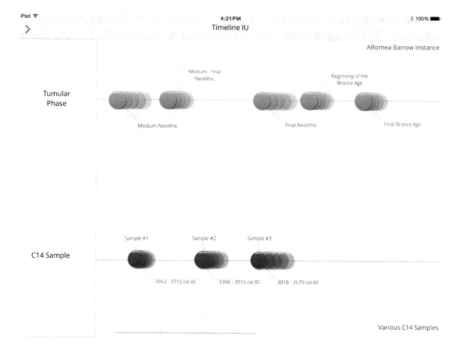

Fig. 9.14 Timeline IU: an example of behavior with two timelines

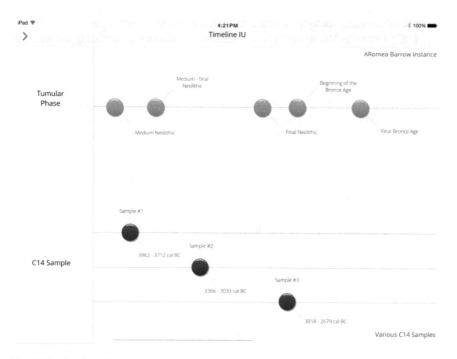

Fig. 9.15 Timeline IU: an example of behavior with two timelines in sub-case 2

The nominal sequence of steps will consist of the archaeologist selecting a dataset and a main class within it along with an instance to be represented. If he/she wishes, he/she can select a second class and an instance from it. Then, he/she will have to decide which temporal attributes are to be represented and, by using the colour control, select the colours he/she wishes to use in order to differentiate the values of those attributes.

The sequence explained above gives rise to two sub-cases. Therefore, the difference in the visualization of these two sub-cases contemplated by the interface should be noted:

- In sub-case 1, the archaeologist chooses the secondary class but not an instance from it: he/she can select a main class and an instance upon which to visualize temporal information. However, he/she can select a secondary class without selecting an instance. In this case, a timeline will be visualized with values of the attributes corresponding to the selected instance which belongs to the primary class and, if an attribute of the secondary class has been selected for visualization, all the values associated to the primary instance will be shown without taking into account to which secondary instance they are associated. In Fig. 9.13, we can observe an example: "C14 Sample Valorization" has been selected as the secondary class but no instance is associated to it; The attribute "C14 Date result" belongs to the "C14 Sample Valorization" class. Therefore, the timeline shows all the samples which exist associated to the instance of the main class (associated to ARomea Barrow), without taking into account to which specific instance of "C14 Sample Valorization" they belong. In order to indicate that they belong to different instances (in this case to different instances of "C14 Sample Valorization") information is maintained regarding the name of the instance (Sample #1, Sample #2). There is the possibility of expanding the attribute's values via an icon under its name (see Fig. 9.15). By selecting this option, each one of its instances and their information is shown separately. This visualization allows for a synthetic view of the attribute associated to the secondary class while maintaining structural coherence and providing the archaeologist with information regarding the number of existing instances belonging to the secondary class.
- In sub-case 2, the archaeologist chooses a secondary class and instance: a timeline is visualized with values of the attributes corresponding to the secondary instance selected, according to the model of the main class. Therefore, the names of the instances in the timeline shall not be shown and neither will there be an icon allowing the timeline to be expanded.

Geographic Area IU

The interaction unit Geographic Area IU deals with problem E (and related with the general challenge 5), identified previously for archaeological data analysis and which makes reference to the need for a presentation and interaction mechanism which allows the geographic facet of the data to be reflected in a way which is adapted to that carried out in archaeology, with its own problems, such as the integration of different systems of geographic location. Therefore, an interface has been designed which allows a main class in a chosen dataset to be selected. This class must have absolute geographic locations associated to it. Optionally, it is possible to select a second class which also has absolute geographic locations associated to it. From these classes, the geographic locations of all the instances present in the dataset corresponding to the selected classes will be uploaded onto the interface. The archaeologist will be able to vary the system of coordinates, the center of the map and the level of zoom in order to adapt the visualization to his/her needs and the characteristics of the screen.

Internal Components

Three individual patterns from level 3 have been selected in order to compose our Geographic Area IU proposal:

- Colour assignment: This pattern has been selected in order to use the colour mechanism in the differentiation of the classes or sub-classes to which the instances visualized in Geographic Area IU belong, allowing the archaeologist to visualize instances of a different nature on the same map.
- Additional Information: This pattern has been selected in order to show additional information about each instance represented, such as the name or the precise geographic coordinates.
- First Focus: This pattern has been selected in order to offer visual clarity in the presentation of additional information about the instance or a set of instances. The archaeologist selects an instance and the interface changes the focus to that element, fading the background in order to highlight the specific element.

Aspects of Use and Interaction

The aim of this interaction unit is to provide the archaeologists and related professionals with a flexible presentation and interaction mechanism, based on the geographic aspect of the data. Figure 9.16 shows a visual prototype. The nominal sequence of steps will consist of the archaeologist selecting a dataset and a main class within it. If he/she wishes, a second class can be selected. Then, he/she will have to adjust the characteristics of the map visualization and, by using the colour

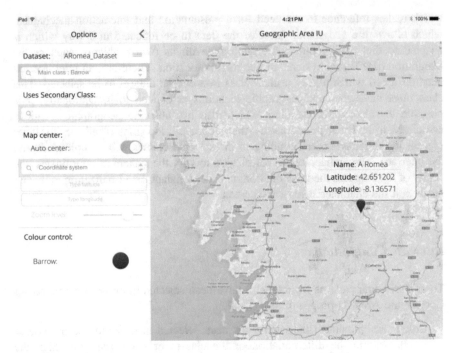

Fig. 9.16 Geographic Area IU: the configuration option can be seen on the left

control, select the colours he/she wishes to use in order to differentiate the values of those attributes. It should be noted that the ultimate aim of this interaction unit is not to serve as an interface for calculation in geographic matters but rather to assist the archaeologist in spatial reasoning. For this reason, the common functions of geographic information systems or similar applications have not been included. Rather, only the functions which allow us to represent the data present in the dataset in its absolute locations and to identify them by their nature have been included. These two components, according to our interviews presented in Chap. 5, are the first questions which the archaeological researcher needs to ask of the data in order to outline his/her analysis of the data, to test research hypotheses and to make decisions based on that data. Later, the complex calculations on geographic matters can be carried out in applications which are more specifically oriented towards geographic calculations.

Sequential IU

The interaction unit Sequential IU deals with problem F (and related with the general challenge 6), identified previously for archaeological data analysis and which makes reference to the need for a presentation and interaction mechanism which allows sequential data or data organized "by levels" to be explicitly included. Therefore, an interface has been designed which allows a main class in a chosen dataset to be selected which will constitute the central element organized by sequential layers. In other words, all the instances belonging to the chosen main class will be visualized in the form of a sequence. Furthermore, the interface will allow the archaeologist to select a second class which must have a direct association with the chosen main class and show additional information about it. Optionally, it will also be possible to visualize in the interaction unit specific aspects such as typologies of the main class or associated temporal phases. The archaeologist will be able to select colour options in order to adapt the visualization offered, due to the fact that it is possible that the sequences presented with several instances will need differentiating elements.

Internal Components

Four individual patterns from level 3 have been selected in order to compose our Sequential IU proposal:

– Colour assignment: This pattern has been selected in order to use the colour mechanism in the differentiation of the temporal phases to which the instances visualized in Sequential IU belong, as well as the valorization groups which the archaeologist forms of the sequential elements.
– Row aggregation: This pattern has been selected in the design of the interface in order to visually organize the sequential levels of the instances presented. Each sequential level is organized into a row, although its limits are not marked on the interface so as not to create visual noise in interfaces with a high number of instances to represent.
– First Focus: This pattern has been selected in order to offer visual clarity in the presentation of additional information about an instance or a set of instances. The archaeologist selects an instance and the interface changes the focus to that element, fading the background in order to highlight the specific element. This is also applicable to sets of instances which function as such in the sequential representation.
– Set: This pattern has been selected in order to represent the groups formed by the archaeologist, which are already stored in the dataset, as elements of the interface with their own entities. In this case, a way of grouping instances (they may be instances belonging to the same or to different sequential levels) has been chosen which superimposes a transparent rectangular element, thus grouping together the instances which fulfil the criterion chosen by the archaeologist.

Aspects of Use and Interaction

The aim of this interaction unit is to provide the archaeologists and related professionals with a flexible presentation and interaction mechanism, based on the sequential aspect of the data. Due to the fact that the nature of the data to be shown sequentially by layers may be extremely different, the decision was taken to define the interaction unit in an abstract way, though illustrating the aspects of its use and interaction with a specific case of application, which requires an analysis by layers based on stratigraphic information. Stratigraphy deals with information relating to the study of archaeological layers or levels of occupation of an archaeological site, and these kind of information are commonly a fundamental source to establish the context of the site and its relative chronology, from which, in turn, the sequence of cultural and temporal evolution is obtained [183]. The selection of this case of application to illustrate Sequential IU can be put down to two fundamental reasons:

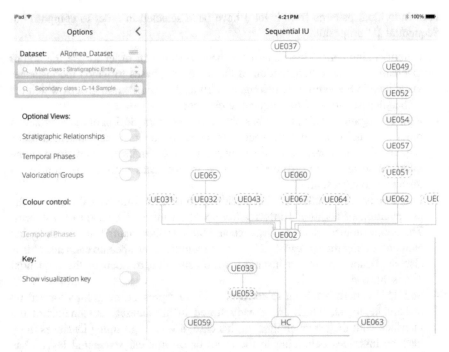

Fig. 9.17 Sequential IU for the visualization of stratigraphic sequences: the configuration options can be seen on the left

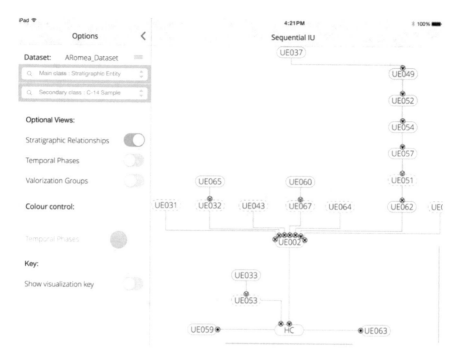

Fig. 9.18 Sequential IU for the visualization of stratigraphic sequences with relations between visualized strata

- The visualization of stratigraphic information in archaeology presents as yet unsolved challenges, which were necessary to deal with from the point of view of interaction and presentation solutions [177].
- The case study presented as the central case study in this book showed necessities regarding the sequential visualization of specific stratigraphic information, which links the scenario of application with the interaction unit selected in order to resolve the problems detected.

Figures from Figs. 9.17, 9.18, 9.19, 9.20 and 9.21 show a visual prototype of the use of Sequential IU for the case of stratigraphic visualization. The nominal sequence of steps will consist of the archaeologist selecting a dataset and a main class within it. The instances of this main class will be those represented sequentially. Therefore, they must contain associated sequential information. In this case of application, it will be a class which conceptually represents the strata of the stratigraphic sequence. If he/she wishes, the archaeologist will be able to select a second class with related information. Then, he/she will have to decide if it is of interest to visualize specific aspects of the main sequenced class, which must be defined ad hoc for each application of Sequential IU, but which abstractly respond

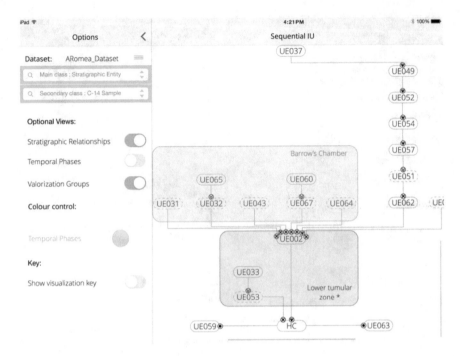

Fig. 9.19 Sequential IU for the visualization of stratigraphic sequences with relations between strata and the evaluations of experts visualized

to typologies within the sequenced elements and to temporal phases associated to them, among other additional information. In this specific case, the archaeologist will be able to select if he/she wishes to visualize typologies of relations between the strata defined within the stratigraphic sequence, temporal phases or evaluations made by experts on those strata or their sequence. These evaluations correspond to groupings of some strata of the sequence carried out by the expert in archaeology. These groupings of strata generally correspond to a scientific interpretation of the function, age, or other aspects of these strata. Furthermore, the colour control allows the archaeologist to select the colours he/she wishes in order to differentiate the associated temporal phases.

Over the course of this section, the different interaction units present in the hierarchy of interaction patterns have been described, along with the individual patterns from level 3 selected for the specific implementation of each interaction unit. The implementation of the hierarchy of patterns proposed as a solution throughout this chapter allows a complete interaction solution to be drawn up for the problems of presentation and interaction of data identified in software assistance for archaeological data sets. It should be noted, however, that the organization of

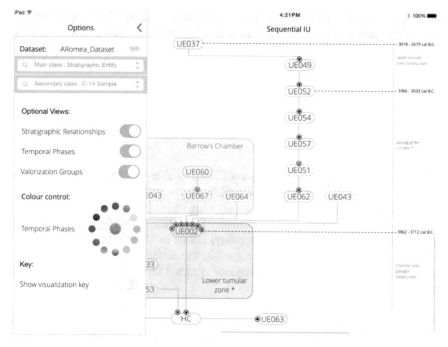

Fig. 9.20 Sequential IU for the visualization of stratigraphic sequences with relations between strata, as well as the associated temporal phases and the evaluations (both made by experts) visualized

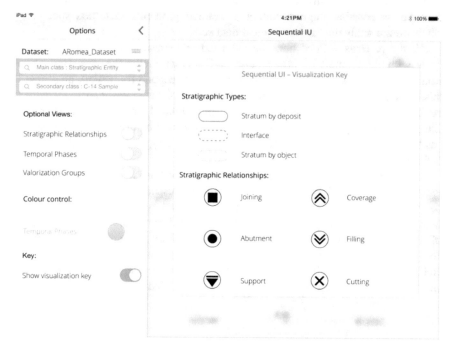

Fig. 9.21 Sequential IU for the visualization of stratigraphic sequences: a view of the key explaining the semantics contained in the different symbols of the diagram

the proposed solution into levels and its possibility of reuse confer upon the proposal described in this chapter a high degree of flexibility, being able to develop alternative designs which deal with other problems (or the same ones with other implementation mechanisms) in archaeology in particular, as well as to redefine the problems and their spheres of application to other fields.

Conclusions

This chapter has presented an analysis of the formal representation techniques for software presentation and interaction which are currently in existence. This overview of existing studies, in addition to allowing us to detect the need (which arose in the context of this research) for formally representing presentation and interaction for software assistance in archaeology, led us to identify series of challenges which are independent of the field regarding the formal representation of presentation and interaction in software assistance contexts for data analysis and decision making.

In this context, a solution based on the pattern concept—more specifically on the definition of a hierarchy of RIA (Rich Internet Applications) patterns—has been proposed oriented towards software assistance to data analysis and decision making. The proposed solution is organized into a three-level hierarchy in order to facilitate its abstract handling and the reuse of the individual solutions proposed. The hierarchy of patterns has been defined in an abstract way, in order to, later describe its possible implementation to resolve problems identified specifically when we are analyzing archaeological data sets.

All the patterns defined here are validated and illustrated in Chap. 11, which deals with the support of presentation and interaction for the case study of A Romea. This case study validates the complete conceptual framework presented in this book and allows an evaluation and validation of the conceptual framework to be carried out with archaeologists and some related professionals. Therefore, Chap. 12 describes how a subset of the interaction solution presented in this chapter has been implemented as a software demonstrator, being empirically validated with archaeologists.

Finally, the innovative contributions presented in this chapter corresponding to the area of the formal representation of software presentation and interaction must be highlighted. On the one hand, a complete proposal, which is independent of the field of application, has been drawn up, based on RIA patterns for the representation of presentation and interaction in applications whose objective is the software analysis of data and decision making. Later, a proposal has been specified for archaeological data analysis, a field in which software presentation and interaction still lacks a corpus of application in data analysis and decision making assistance. In addition, the specific proposal of implementation of the interaction unit Sequential IU in archaeology constitutes, as has been seen throughout the study of previous research [179] and during the chapter, a novel solution for stratigraphic visualization, which is cognitively adapted to the needs detected by archaeologists. Thus highlighting the resolution of problems which emerge regarding the non-separation between the material dimension and the interpretative dimension of the stratigraphic information, a problem which was detected in our previous research [177].

Chapter 10
Integration, Interoperability and Consistency Between Framework Models

Introduction

A software system is not a conceptually one-dimensional artefact, whatever the purpose for which the software was conceived or the (in our case archaeological) data was collected. The development of any software system implies that its different aspects work in a horizontal space, a fact which must be taken into account. These aspects are parts of a common metamodel of the system, which requires specific models in order to define (or prescribe) how the system supports the proposed requirements, fulfilling the objectives defined for it. Therefore, a conceptual framework in three parts (subject matter, cognitive processes and presentation and interaction mechanisms) has been proposed as a solution. This proposal has given rise to three conceptual metamodels which reflect these three aspects of the framework in an independent manner. In order to achieve the established objective, the framework must have an internal consistency, allowing the relations which exist between the three parts identified to provide assistance in this way to the archaeologists to be expressed. This makes it necessary to formally define the connections between the models, as well as the possibilities for interoperability which exist between the three models defined.

Thus, it is necessary to define an integration mechanism which enables interoperability to these perspectives in order to compose a unique conceptual model of the system. This mechanism will provide a specific reference to deal with software creation tasks directed by models (MDE), such as automatic compilation, the verification of models, the evolution of models, the execution of metrics and model analysis operations, etc. The proposal outlined here deals with the specific field concerning here: software systems created in order to assist archaeologists and related professionals to the generation of knowledge in archaeological data environments. More specifically, our framework focuses this assistance on offering and applying adapted visualization techniques according to the model defined by Chen [61] integrated in our proposal (see chapters of Part III), and taking into account the

© Springer International Publishing AG 2018
P. Martin-Rodilla, *Digging into Software Knowledge Generation in Cultural Heritage*, Modeling and Optimization in Science and Technologies 11, https://doi.org/10.1007/978-3-319-69188-6_10

implications on knowledge extraction techniques which this assistance may have. Due to the growth in the application of software-assisted knowledge generation systems in scientific disciplines, it is necessary to deal with the formal expression of the different modeling perspectives which are present in these systems. The expression of the relations between models allows the generation of scientific knowledge to be managed at a high level of abstraction, which implies fully dealing with the software-assisted knowledge generation process. For example, it is possible to capture the relation between available scientific data, the information or results produced and the cognitive processes involved in producing them. If we do not take into account these relations, it is extremely difficult to obtain data regarding traceability in this process or to replicate each research in order to validate or reuse the work done.

With this objective in mind, this chapter analyses the necessities as far as internal consistency and interoperability of software assistance systems for the generation of knowledge are concerned. Later in the chapter, an interoperability framework [211] will be considered in order to propose a specific metamodel permitting the integration of different modeling perspectives which intervene in our framework proposed. Hence, the chapter treats their conceptual and technical details. If the reader is only interested in the directly application of the framework to the archaeological data sets, it should be noted that the integration metamodel defined will later be used in a real scenario: the specification of relations between models of the three different perspectives of the proposed framework in this book for archaeological data sets, being validated by their implementation and application in a case study in Chaps. 11 and 12.

Existing Works on Expressing Connections Between Models

As was defined in Chap. 2, knowledge generation has been defined by many authors as a model formed by levels or layers from the raw data to levels of higher abstraction (knowledge, wisdom, etc.). In all of these models, the transition from one level to the next passes through the application of cognitive processes. Over the course of previous chapters, we have defined how the field which concerns us and the subject matter to which the data we have belongs can be conceptually expressed (Chap. 7), how to deal with and characterize the cognitive processes which take place in the generation of knowledge (Chap. 8) and, finally, how to formally express the data presentation and interaction patterns which favor and assist to the generation of knowledge in archaeology (Chap. 9).

The three modeling perspectives mentioned above constitute the three parts of our framework and must be correctly integrated with the aim of offering assistance to users in the generation of knowledge. This means that we can formally express the relations between the subject matter in which knowledge will be generated and cognitive processes, being able to integrate scientific data with the reasoning processes involved in the generation of knowledge. In the same way, we can formally express the relations between the aforementioned subject matter, the cognitive processes and the recommended mechanisms for presentation and interaction. Thus, it is possible to recognize, for example, that certain scientific results have been obtained by comparing (having carried out the cognitive process) two sets of data and/or specific values and to offer adapted visualization patterns.

Within the field of Software Engineering, there are several approaches for the modeling of the different perspectives (or their combinations) in a software system. For example, approaches based on requirements [16, 77, 279], related with the data or the field it concerns [149] or approaches oriented towards objects [198, 210, 212]. In this context, it is in the field of Model-Directed Engineering (MDE) [11, 29, 68] where, over the course of recent years, the need to formally express relations between models regarding different views of the same software system has been identified. There are different techniques to express this relation between models: model weavings, model mappings, pivot metamodels and pivot ontologies, among others. All of these techniques have been applied successfully, especially in the reduction of the conceptual distance between the definition of the business areas of large companies and their related software systems [243, 262] or in complex integrations of requirements and functional models [168, 292]. We can also find some interoperability studies focused on software development for software-assisted knowledge generation [92]. However, the providing of software assistance to knowledge generation has not been dealt with from an interoperability perspective based on models (model-driven interoperability). The formal expression of the relations between the models which take part in the process of software systems development is one of the main objectives of existing interoperability techniques. Therefore, we believe that this perspective allows the relations between models in the specific case of software-assisted knowledge generation to be expressed.

In this context, we have adopted the definition of "interoperability" as "the ability of two or more systems or components to exchange information and to use the information that has been exchanged" [226]. When the information exchanged is represented by models, we refer to interoperability based on models (model-driven interoperability). Thus, the exchanged information contains the relations between the different perspectives involved in a specification of the software system. In other words, models are used to represent the different points of view and their connections via interoperability techniques. There are numerous different techniques and approaches to formally express bidirectional relations between models.

Taking the existing literature on model-driven interoperability as a basis, we have identified some studies of special relevance which deal with the expression of relations between models representing different perspectives:

The approach taken by El Hamlaoui [73], for example, expresses the connections between models with a metamodel of correspondences. However, this approach presents a multi-user definition. In other words, different users interact with different perspectives of the same software system. In our case, though, the same user profile interacts with all the perspectives defined in the system. For example, several researchers (though always with a single user profile) interact with all the perspectives of the software assistance system in order to carry out knowledge generation tasks. Other solutions in existence, such as the Capella tool [221]—based on the Arcadia method [79]—, express the connections between models from an interoperability point of view directed by models. However, this approach is designed to provide support for industrial processes and not for assistance to the generation of knowledge. More examples of studies on interoperability can be found in the detailed review carried out by Pastor [211].

Furthermore, a significant number of authors have highlighted the need for a combination of techniques with the aim of improving results or confronting the different needs which arise. Some good examples include the uses of UML profiles for modeling specific perspectives of a system [243], the use of ad hoc processes based on diagrams of characteristics in order to find and represent the relations between models [287] and examples of the application of combined interoperability techniques with the aim of transforming one model to another [67, 262]. We can also find similar examples applied to the field of knowledge management, expressing interoperability between databases [232].

As can be seen in all these studies, there are many techniques to express the relations between models of different perspectives from an approach based on model-directed interoperability. However, they have not been applied to the field of software-assisted knowledge generation. Due to the existence of an abstract framework [211], which allows us to evaluate the different techniques and artefacts of interoperability (independently of the specific techniques and technology applied), our proposal consists of taking this framework as a reference point in order to design a mechanism which allows relations between models in software-assisted knowledge generation systems to be formally expressed. In our proposal, Chen's proposal [61] is adopted as the method of software assistance, offering adapted visualization patterns.

The following sections describe the work carried out in this respect, presenting the different parts of the framework in terms of interoperability.

Our Approach Based on Interoperability

As has been seen in previous chapters and following Chen's foundation work [61], the assistance given to users in the generation of knowledge is materialized in our framework in recommendations through adapted presentation and interaction patterns. According to that, there are three important aspects in any software-assisted knowledge generation system: the subject matter upon which the users will generate knowledge, the cognitive processes carried out by these users and the adapted presentation and interaction proposals which the system must offer. In order to express these three aspects, different modeling solutions can be found, and three metamodels have been created to represent these aspects: the subject model (named SM in this chapter), the cognitive processes model (named PCM in this chapter) and the interaction model (named IntM in this chapter).

As has been mentioned in previous chapters, the subject model of the framework corresponds to the ConML metamodel [129], a metamodel which, despite not having limitations in definition about the field of application, was specifically conceived in order to express conceptual structure in matters of Humanities and Social Sciences, although our interest is focused on archaeological domain. However, it should be noted that it would be possible for the subject metamodel of the framework to be any other metamodel oriented towards objects (UML for example) in the terms of interoperability which concern us here. Metamodels of cognitive processes and of interaction do not have limitations as far as the field is concerned. In other words, they capture the semantics necessary for the expression of aspects regarding cognitive processes and the possibilities for presentation and interaction at a high level of abstraction, which allows them to be applied to any software-assisted knowledge generation system which we may need to design.

In order to obtain an integrated model which considers the different perspectives proposed in this research (subject matter, cognitive processes and presentation and interaction mechanisms), we have applied the MDD interoperability model defined by Pastor [211]. This model establishes interoperability based on models in three dimensions: semantic, syntactic and technical interoperability.

Semantic interoperability refers to the semantics belonging to the modeling approaches which we wish to interoperate with and is normally specified in textual representations (such as in UML [203] or i*specifications [4, 289]). In Pastor's MDD approach, it is considered that the semantics are implicit in the connections defined between the conceptual constructs which the different modeling perspectives to interoperate with represent. In addition, there is a warning regarding the lack of a standard for the definition of this semantic interoperability. Taking this into account, we describe the semantic interoperability by way of the models corresponding to the three dimensions being dealt with, as can be seen in Figs. 10.1, 10.2 and 10.3.

Syntactic interoperability (abstract syntax) refers to the particular system of representation of the semantics described above. In this case, this syntax is obtained through the use of a common metamodeling language (Essential Meta-Object Facility, commonly abbreviated EMOF) [202] for all the modeling approaches involved. Technical interoperability refers to the format used for the exchange of information between dimensions being interoperated with and, in this case, is achieved via the XML specification provided by the Eclipse tools [87] used (UML2 y EMF).

On the other hand, Pastor's framework establishes that, for the implementation of any interoperability artefact or for the automatization of the operations based on the models involved, it is necessary to specify additional aspects, in terms of Procedure, Application, Infrastructure and Support for the representation of data.

The Procedure aspects make reference to the elements which must be defined, as well as to the steps which must be taken in order to achieve the sought-for interoperability. The procedure used in order to integrate the different perspectives

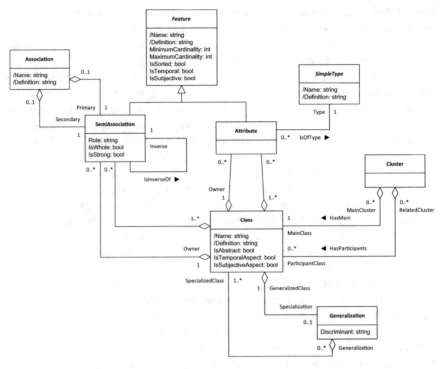

Fig. 10.1 ConML Metamodel, playing the role of Subject Model (SM throughout this chapter)

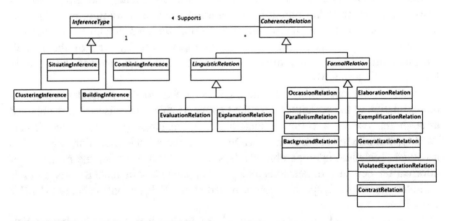

Fig. 10.2 Cognitive Processes Metamodel (PCM throughout this chapter)

involved in this case consists of three steps: (1) the definition of the metamodels; (2) the identification of equivalences and differences between them and (3) the definition of an integration metamodel which acts as a pivot solution for the representation of equivalences between metamodels, the new information generated

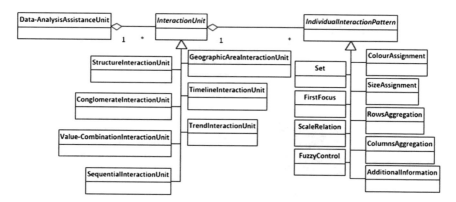

Fig. 10.3 Interaction Metamodel (IntM throughout this chapter)

necessary for the later application of the integration metamodel and the handling of heterogeneities on a modeling level between the three perspectives of data, cognitive processes and presentation and interaction mechanisms.

The Application aspect defines the utilities or tools used in order to achieve interoperability. In this case, the procedure described above has been implemented based on the utilities provided by Eclipse EMF (Foundation 2014), which are used for the generation of a model editor on the basis of a specification of an Eclipse UML2 metamodel.

These applications also provide the definition for aspects corresponding to Infrastructure and Support for the representation of data. The Infrastructure aspect corresponds to the definition of the mechanisms of communication between the applications in order to ensure the correct exchange of information and in order to avoid the loss of modeling information when this exchange process is carried out. In our case, the defined infrastructure has been the use of XML as the exchange format.

Finally, the aspect corresponding to the Support for the representation of data deals with the format in which the modeling artefacts are found and must be specified in a standard format, which can be interpreted by different modeling tools independently of application platforms and contexts of development. In our case, we have opted for an EMF representation for the resulting modeling artefacts.

The Integration Metamodel

Here, we shall describe the integration metamodel proposed for the formal expression of the relations between the models of the framework in terms of model-driven interoperability. The full integration metamodel can be seen in Fig. 10.4:

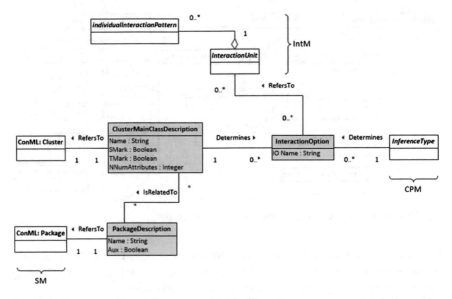

Fig. 10.4 The proposed integration metamodel. The classes in grey correspond to the integration model. The classes in White are part of the metamodels shown in the three previous figures

The integration metamodel presents three classes: *ClusterMainClassDescription*, *PackageDescription* and *InteractionOption*.

The *PackageDescription* and *ClusterMainClassDescription* classes encapsulate all the necessary semantics which the metamodel of data establishes as far as the structure and meaning of the data involved is concerned. *PackageDescription* consists of two attributes:

- *Name*: Attribute of a text/chain type. Its value corresponds to the name of the class which plays the role of the main class of the package.
- *Aux*: Attribute of a Boolean type. It will present a True value if the package selected forms part of/plays the role of auxiliary for the rest of the packages within the subject extension with which we are working and False in the opposite case.

The *ClusterMainClassDescription* class consists of 4 attributes:

- *Name*: Attribute of a text/chain type. Its value corresponds to the name of the class which plays the role of the main class of the cluster.
- *SMark*: Attribute of a Boolean type. Its value indicates if the main class of the cluster is associated by way of associations with a subjective mark with other classes in the cluster.
- *TMark*: Attribute of a Boolean type. Its value indicates if the main class of the cluster is associated by way of associations with a temporal mark with other classes in the cluster.

- *NNumAttributes*: Attribute of a number data type. Its value corresponds to the number of numerical attributes (enumerated or Boolean) which are present in the main class of the cluster. This information will allow us to then make decisions with regard to these attributes.

Furthermore, the relation *IsRelatedTo* between both classes allows it to be known which classes act within which package. This information is already indirectly present in the ConML metamodel since we know which classes form part of which packages and which classes form part of which clusters. However, this relation in the integration metamodel allows the relation to be made explicit. By evaluating the information contained in the instances of *PackageDescription* and *ClusterMainClassDescription* and the structure of instances arising in each case for *SM*, *CPM* and *IntM*, the integration metamodel allows us to determine which interaction options (via instances of the *InteractionOption* class) are appropriate in each case. The *InteractionOption* class consists of a sole attribute, *IOName*, of a text/chain type, which gives a name to each interaction option instanced. This last part shows software assistance by way of recommendations of presentation and interaction patterns modelled in our framework.

Interoperability Guidelines

As can be seen in Fig. 10.4, the integration metamodel captures the three perspectives which participate in any software-assisted knowledge generation system, with the aim of offering the end user more appropriate visualization techniques to assist him/her in his/her tasks [61]. The decision regarding which specific interaction unit is most appropriate according to the cluster of data being dealt with and the cognitive process being assisted is a choice which the software analyst or domain specialist should make when it comes to implementing the framework presented here. However, we believe that there may be some universal guidelines which we have observed in the scenarios tested with our interoperability model [175]:

- Each cluster defined and each inference of a Building Inference type can determine an interaction option which refers to a Structure IU interaction unit. This interaction unit enables assistance to problems, such as Problem 1 (which is defined on an interaction level), related with the necessity of exploring the internal structure of the dataset of the case study.
- Each cluster defined and each inference of a Clustering Inference type can determine two interaction options:

    (1) An interaction unit of a Conglomerate IU type, which allows the values associated to instances belonging to the cluster's main class to be explored. This interaction unit allows us to assist to problems such as Problem 2,

(which is defined on an interaction level), which are related with the necessity for dynamism in the visualization of datasets.

(2) An interaction unit of a Value-Combination IU type, which allows the values associated to attributes and to classes belonging to auxiliary packages which are associated to the cluster's main class to be explored. This inter-action unit allows us to assist to problems such as Problem 2 (which is defined on an interaction level), which are related with the necessity for dealing with levels of importance throughout the visualization, obtaining, if desired, additional information on a piece of evidence, an event or a reference or a subset, whilst maintaining on the screen the general view of the cluster being dealt with.

• Each cluster defined and each inference of a Combining Inference type can determine two interaction options:

(1) An interaction unit of a Value-Combination IU type, which allows the values associated to attributes and to classes belonging to auxiliary packages which are associated to the cluster's main class to be explored. This inter-action unit allows us to assist to problems such as Problem 3 (which is defined on an interaction level), which are related with the necessity for dealing with levels of importance throughout the visualization, obtaining, if desired, additional information on a piece of evidence, an event or a reference or a subset, whilst maintaining on the screen the general view of the cluster being dealt with.

(2) An interaction unit of a Trend IU type in cases in which the value of the NNumAttributes attribute of the CusterMainClassDescription class is higher than 2 (the more attributes of this type, the more appropriate the Trend IU interaction unit will be). This implies that the main class of the cluster dealt with presents various numerical, enumerated or Boolean attributes, which are potentially susceptible to varying considerably from one version of the dataset to another. This interaction unit allows us to assist to problems such as Problem 4 (which is defined on an interaction level), which are related with the necessity for carrying out analyses of similarities or differences between the data at different stages of the research.

It should be noted that these guidelines take on fundamentally structural aspects when it comes to deciding which interaction unit is more appropriate compared with others when presented with a data cluster and an inference to be assisted. Due to this fact, they are independent of the field of application of the framework and, therefore, it is desirable a careful analysis on whether they are appropriate for other fields (outside of the field of archaeology) which are not explored in this work.

Other application guidelines, of a semantic nature, do depend strongly on the field of application in which we wish to provide assistance. The guidelines extracted for archaeology are described in Part IV, in which a complete imple-mentation of the framework for assistance to the generation of knowledge in archaeology is carried out.

Finally, it must be highlighted that, in order to obtain a full representation of the relations implied in any software-assisted knowledge generation system, it is necessary to instance all of the metamodels, including the integration metamodel, which acts as a pivot. Over the course of following Chap. 11, a description is given of the application of the integration metamodel created for our real-life scenario: software-assisted knowledge generation systems in archaeological environments. This application is carried out by way of a case study selected for the analytical validation of the complete framework and includes an implementation of the integration metamodel presented in this chapter.

The formal expression of the relations between the different perspectives of the framework, which are obtained thanks to the integration metamodel presented here, will allow us to guarantee the traceability, reuse and replication of studies carried out in archaeology which use software-assisted knowledge generation systems.

Conclusions

To sum up, this chapter presents a solution, by way of an integration metamodel, in order to formally express the relations between metamodels in software-assisted knowledge generation systems. In addition, some universal guidelines have been described which allow the software analyst or domain specialist to approach software-assisted knowledge generation from the point of view of adapted presentation and interaction patterns. Although the interoperability techniques applied in the definition, design and later implementation and validation of the integration metamodel are not novel, their application in order to express the relations which exist between models in software-assisted knowledge generation systems does represent a scientific contribution. It is an application which, to our knowledge at the time of writing this book, has never before been employed. Furthermore, no studies have been found which, in matters related to knowledge generation systems, take into account the cognitive processes carried out by the users of a system as a perspective of the system which must be present in the integration metamodel.

The framework presented acts, in this real-life scenario, to carry out an analytical validation in archaeology of the integration metamodel presented here and described in Chap. 11. The empirical validation of the proposal with archaeologists allows us to discover and show the readers the degree of added value which the formal integration of the framework's perspectives offers to the end users, as well as to obtain *feedback* in order to make improvements in the future. All of this is also described in Chap. 12.

Part IV
Archaeological Applications in Practice

Chapter 11
Analytical Validation: A Romea as a Case Study

Introduction

At this stage of our research, it's time to put what we developed and learned into practice applying it to a real case study.

The validation of theoretical proposals by way of a real case study has been, and continues to be, a constant within the validation mechanisms employed according to hypothetico-deductive methodologies, such as Design Science Methodology [281] or other similar methods. Research projects validated by way of case studies can be found in many different fields [83], from Sociology [280] or Psychology [239] to Medicine [3], Education [184], etc. In Software Engineering, this is a widely-used validation mechanism, with case studies being found in validations of requirements models, functional models, testing, empirical validations of business systems or processes [281, 282], etc.

This chapter we shall present the analytical validation which has been carried out of this proposed framework via a real archaeological case study: A Romea. Thus, we show how the software metamodels of the defined framework allow represent all the information regarding the three aspects necessary in order to provide assistance to the generation of knowledge (subject matter, cognitive processes and presentation and interaction mechanisms), and the degree of interoperability between them, in a real archaeological case study and how we can define the software assistance offered in this case. The reader will be able to follow step by step the application of the framework defined to the archeological data sets and particularities of the case study, in order to server as a guide for future applications to their own cases. In addition, A Romea serves as a method for detecting possible problems in its application and possible lines for improvement, determining a specific context of application and allowing us to adopt an appropriate approach as far as the generalization of the solution to other similar contexts is concerned.

© Springer International Publishing AG 2018
P. Martin-Rodilla, *Digging into Software Knowledge Generation in Cultural Heritage*, Modeling and Optimization in Science and Technologies 11, https://doi.org/10.1007/978-3-319-69188-6_11

A Romea as a Case Study: A General Overview

In this section, the particularities of the A Romea case have been explain in detail. It is possible that, for most of our readers, the terminology and definition of archaeological entities sounds repetitive or excessively familiar. However, it is necessary to define adequately the scope and terminology of our case study to apply the solution later, being also a recommendation when we apply the framework to our own cases, archaeological datasets, etc.

The case study selected forms part of a group of archaeological campaigns carried out with the aim of mitigating the impact on archaeological and Cultural Heritage entities caused by the construction of the Santiago-Alto de Santo Domingo motorway in Galicia (North-West Spain) [56]. More specifically, during the work to assess the archaeological impact of the construction of this motorway carried out between December 1999 and February 2000, an underground spatial concentration of material remains of human activity (known as the Barrow of Monte da Romea) [169, 225] was documented on the plotted route of the motorway. This, in archaeological terminology, was identified as an unregistered site [183].

Traditionally, "barrow" is the name given to an artificial mount of earth and/or stones heaped up to cover one or more burials [64]. Barrows were common throughout the course of Prehistory in different cultural contexts [183]. The corrective measures proposed in order to mitigate the destruction of the barrow consisted of excavating the site before the construction of the motorway was carried out, in order to obtain as much information about it as possible. Archaeological excavation, as a fundamental method for obtaining archaeological data [64], consists of discovering and registering material remains of past human activity [183]. This research procedure uses appropriate methods to unearth these remains, which may include artefacts, organic remains and structures [183]. All of the elements found, with the obvious exception of structures (such as walls, etc.), are transferred to a laboratory in order to be analyzed, thus generating more data. The excavation is carried out by extracting the earth which covers the remains, following natural or artificial layers (strata), which make up the stratigraphic sequence of the excavated site, of a thickness which is decided upon by the archaeologist [64, 183]. As excavation is an irreversible act, only those sites which can potentially provide new information or which are in danger of destruction [183] should be excavated, as was the case of A Romea [56].

The excavation work was carried out between January 2001 and January 2003, in parallel with the construction phase of the motorway, by an independent archaeological team [56].

During the project, the existence of the barrow, its location and the evaluation of the impact suffered in the previous phase was confirmed. The barrow was located totally within the bounds of the construction work, thus making the work critical as the construction of the motorway would suppose the destruction of the site. Confronted with this situation, a zone of archaeological caution was established (a geographically delimited zone in which no movement of the earth, or of machinery, is permitted). In addition, some corrective measures were proposed:

- Graphic and cartographic documentation of the area.
- The intensive prospection of the area. By archaeological prospection, we understand the application of a set of methods in order to discover sites based on their superficial remains by visual inspection on the ground or from the air (aerial photography, teledetection, etc.), or from remains unearthed close to the surface by employing apparatus which measures chemical, electrical or magnetic variations in the soil (geophysical prospection) [64, 183].
- The marking out of the area during the work.
- The archaeological control of the clearing of the land.
- The mechanical digging of trenches/ditches within the limits of the barrow.
- The complete excavation of the site before the construction work was carried out at this geographic location.

The carrying out of these corrective measures gave rise to the project which is taken here as a case study and which is identified by the code CJ102A 2002/100-0, according to the appropriate authorization of the Dirección Xeral de Patrimonio Cultural (DXPC) (General Directorate of Cultural Heritage) of the Xunta de Galicia (the regional parliament in Resolution of 25th March 2002).

Description of the Site and the Associated Archaeological Research

The barrow of A Romea is situated in the council of Lalín in an area of inland valleys (see location in Figs. 11.1 and 11.2), in the district of Trasdeza, in the province of Pontevedra (Spain). Prior to the excavation, the visible barrow was covered by thick vegetation and by a variety of repopulated (pine) and autochthonous (oak) trees. It was 21 m in diameter and one meter in height with an oval shape and steeper slopes on the south (S) and east (E) sides and more gradual slopes towards the west (W) and north (N). Once the vegetation had been cleared and the barrow cleaned, it measured 18.90 m in length from north to south (N-S) and 18.50 m from east to west (E-W), reaching a maximum height compared to its surroundings of 1.25 m along its south-east edge and a minimum height of 0.60 m along its northern (N) edge.

Taking into account these initial aspects of the project mentioned above, some archaeological and heritage objectives were proposed for the investigation of the A Romea site (which constitute the overall objectives which the framework proposed here must achieve through software assistance). More specifically, the following objectives were proposed on an archaeological level [56]:

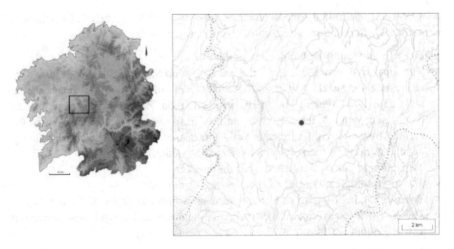

Fig. 11.1 The geographic location of the site of A Romea. Figure provided by Patricia Mañana-Borrazás (©Incipit, CSIC)

Fig. 11.2 A topographic plan of the A Romea site. The red lines indicate the contours at different levels of elevation of the barrow. Figure provided by Patricia Mañana-Borrazás (©Incipit, CSIC)

1. An approach towards a more exhaustive characterization of the site and of the area in which it is located.
2. A verification of the existence or absence of associated archaeological remains in the surrounding area.

3. A definition of the morphology of the site and the remains found and the proposal of a hypothesis regarding its origin.
4. A full and detailed reading of the stratigraphic sequence paying attention to both formal and archaeological aspects. The relation between the characteristic material remains of the site and the different stratigraphic layers should be determined.
5. Samples should be obtained allowing for an analysis which will make it possible to clarify more exactly the chronology and configuration of the site and the structures excavated.

Therefore, we can assume that the archaeological research questions to be answered (and assisted) will be:

- Do the archaeological findings confirm the existence of a barrow? What is the morphological structure of the mound? What is its oldest chronological adscription?
- What later phases can be attributed to the barrow (phases of occupation, abandonment, change of use, etc.)? Do the material findings correspond to the stratigraphic sequence of the site?
- Are there any nearby areas of activity generated by the presence of groups of humans linked to the barrow during its construction or even before?

To sum up, the fundamental objective on an archaeological and heritage level has two sides. On the one hand, there is a need to exhaustively document all the archaeological findings (be they objects, structures such as the stratigraphic sequence or the surrounding area), due to the site's imminent destruction for the construction of the planned motorway. On the other hand, there is the need to provide answers to the research questions mentioned above based on the data gathered during the excavation and which could be summarized by tracing the chronological phases and the activity of the barrow, based on data regarding the findings (objects, structures or the stratigraphic sequence). The answers to these questions regarding the specific case study are those which will be assisted in the proposed framework.

It should be remembered that, as has been emphasized throughout this research, during the phase of exploring the problem, no integral proposal for software-assisted knowledge generation in archaeology has been found in the current literature. What was found were isolated proposals for handling data or for the visualization of subsets of information. The absence of a proposal and the revision of the proposed case study allowed us to ensure that the answers to these research questions at the time the project was carried out (years 2001–2003) were obtained without an integral software assistance system involved in the process of knowledge generation, although some assistance was provided in some cases in terms of spatial information. As will be discussed later in this chapter, a case study has been chosen in which the answers to the research questions and the knowledge generated from the data can be empirically evaluated in pre-software assistance and post-software assistance contexts with the framework proposed, thus allowing us to make comparisons between these two situations.

Therefore, the A Romea project serves well as a case study in the validation of the different solutions which make up the framework presented in this book. The case study acts, therefore, as a central thread of this research and illustrates how software assistance can be achieved by applying the proposed framework and what tangible results can be obtained thanks to its application within a real-life archaeological scenario. The following sections of this chapter go on to describe the specific applications of the three perspectives of the framework as applied to the case study of A Romea.

A Romea: Subject Model

As has been mentioned in the general overview of our case study, in an archaeological excavation, a series of artefacts, biological remains and/or structures are extracted and documented and their associated information serves as a basis for the archaeologists to infer and generate knowledge regarding facts from the past. More specifically, the archaeologist can ascend through the DIWK hierarchy [7], advancing from the data to more complex forms of knowledge. In each case, the type of knowledge which is generated will depend on the research questions which the archaeologist, or a group of specialists, proposes. In this specific case, the knowledge generated from the data will be that which provides answers to the research questions mentioned above. This knowledge, therefore, can be summarized as:

- New knowledge regarding evidence which confirms the existence of the barrow, its internal morphological structure, its possible functional organization and its oldest chronological adscription.
- New knowledge regarding the temporal phases associated to the barrow and its relation with the temporal phases associated to the materials found within it.
- New knowledge regarding the areas of activity which existed in the area surrounding the barrow.

In order to provide software assistance to this process of knowledge generation, it is necessary to possess the data extracted during the excavation process and during the laboratory analysis of the materials found. However, as is mentioned in Chap. 7, the underlying conceptual model which we use for the conceptualization and storage of this data plays an extremely relevant role in the generation of knowledge which can be accessed later and, of course, also in the type of software assistance which can be provided in this generation of knowledge. Bearing these aspects in mind, a conceptual model has been designed which reflects the subject matter dealt with in this case study as an extension of CHARM [99] (as a consequence of this it is expressed in ConML [129]).

In Fig. 11.3 the extension created for the case study of A Romea is shown. The classes and associations shown in orange represent classes and associations which are specific to the extension of CHARM created. The classes in green represent CHARM classes which have been redefined for the case study of A Romea.

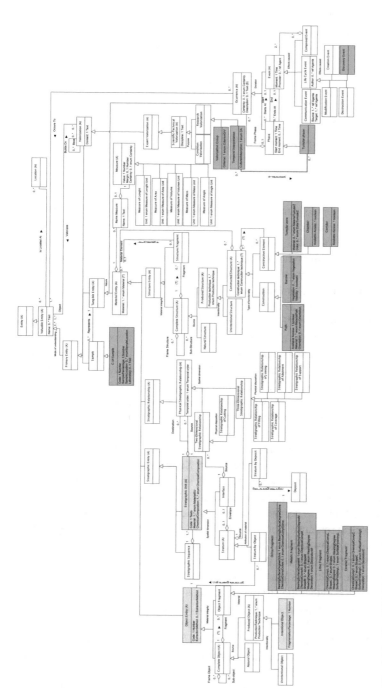

Fig. 11.3 The CHARM extension (based on version 0.8 of CHARM and expressed in ConML) made for the case of A Romea. The classes and associations in orange represent classes and associations which are particular to the CHARM extension created. The classes in green represent CHARM classes which have been redefined for the case study of A Romea

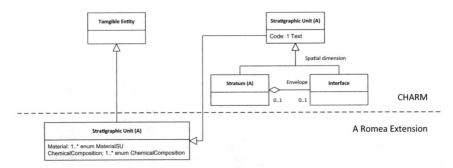

Fig. 11.4 Redefinition technique applied for green classes

The extension presented here covers the complete subject sphere of the case study: the structure and characteristics of the site in question, the types of findings and their specific characteristics, spatial and temporal information regarding the site and its findings, the stratigraphic sequence defined during the excavation process and data obtained in the laboratory. Furthermore, it includes the different evaluations made by the team members who participated in the A Romea project regarding the entities mentioned above.

In its current version, ConML does not allow the in-place redefinition of classes. To achieve a similar effect, the mechanism shown in Fig. 11.4 is employed. A class is added as a specialization of a CHARM class, and used as a generalized class from the CHARM class that we want to redefine. Figure 11.4 illustrates the redefinition technique used in the Stratigraphic Unit class redefinition. The same technique has been applied to all classes from CHARM redefined for A Romea extension— classes in green in all figures in this chapter.

In addition, the framework presented organizes the defined extension of CHARM into subject sub-areas, by using the package mechanism. In this case study, the extension has been organized into 4 packages: Stratigraphic Unit, Object Entity, Structure Entity and Valuable Entity. In turn, the Valuable Entity package is divided into 5 sub-packages: Sample, Measure, Location, Occurrence and Valorization. It should be noted that, as was defined in Chap. 7, each package takes the name of the class with the greatest semantic load within the package and which brings together its subject matter with a view to using the package mechanism as a subject aggregator in the integration model defined for the framework (see Chap. 10).

The Stratigraphic Unit package includes the classes of the extension model for A Romea which conceptualize the stratigraphic units and the relations between them. Thus, the Stratigraphic Unit class, which gives its name to the defined package, represents the class with the greatest semantic load within it. The package also has a redefinition of the CHARM class *Stratigraphic Unit*, incorporating new attributes of use into it for the case study in question. The complete Stratigraphic Unit package can be seen in Fig. 11.5.

The Object Entity package includes the classes of the extension model for A Romea which conceptualize the objects found during the archaeological activity. Thus, the Object Entity class, which gives its name to the defined package,

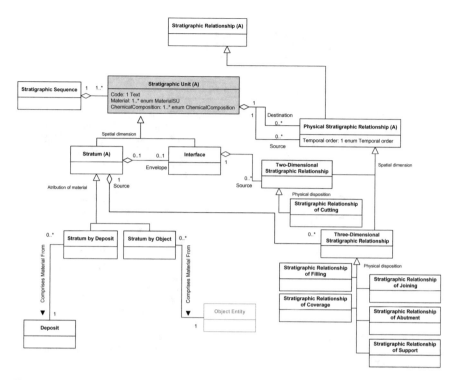

Fig. 11.5 The stratigraphic unit package defined for the case study of A Romea. The object entity class is not part of the package but is shown in order to indicate its association with the classes of the stratigraphic unit package

represents the class with the greatest semantic load within it. In this package, four specialized classes of Object Fragment are included as a result of the extension made: Ochre Fragment, Metallic Fragment, Lithic Fragment and Ceramic Fragment. The definitions of the extended classes are as follows:

- Ochre Fragment: An object entity corresponding to a separate portion of a complete object, having an altered material integrity, and whose material consists of a natural earth pigment containing hydrated iron oxide, which ranges in colour from yellow to deep orange or brown.
- Metallic Fragment: An object entity corresponding to a separate portion of a complete object, having an altered material integrity, and whose material consists of chemical elements that are typically hard, opaque, shiny, and has good electrical and thermal conductivity.
- Lithic Fragment: An object entity corresponding to a separate portion of a complete object, having an altered material integrity, and whose material is mainly composed of stone.
- Ceramic Fragment: An object entity corresponding to a separate portion of a complete object, having an altered material integrity, and whose material is inorganic, non-metallic solid comprising metal, non-metal or metalloid atoms primarily held in ionic and covalent bonds.

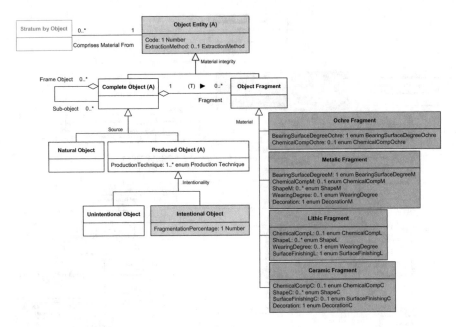

Fig. 11.6 The Object entity package defined for the case study of A Romea. The stratum by object class does not form part of the package but is shown in order to indicate its association with classes of the Object entity package

Therefore, it has been necessary to redefine the CHARM classes *Object Entity* and *Intentional Object* by incorporating new attributes of use for each case study dealt with. The complete Object Entity package can be seen in Fig. 11.6.

The Structure Entity package includes the classes of the extension model for A Romea which conceptualize the structures found during the archaeological activity. Thus, the Structure Entity class, which gives its name to the defined package, represents the class with the greatest semantic load within it. In this package, specialized classes have been added in order to represent specific structures and their fragments: Path, Barrow, Tumular Zone, Chamber and Corridor. The definitions of the extended classes are as follows:

- Path: A constructed entity made to facilitate the movement of people, animals or vehicles, such as a road, way, or track.
- Barrow: An artificial accumulation of dirt and/or rocks forming an artificial mound, especially over a grave, frequent over the course of Prehistory in different cultural contexts.
- Tumular Zone: A constructed entity which, despite not providing direct functionality to its users, constitutes a material part of a barrow, to which it contributes structure and/or function.
- Chamber: A constructed entity which constitutes a material part of a barrow. It's a room built from rock or sometimes wood, which could also serve as a place for storage of the dead from a family or social group and was often used over long periods for multiple burials.

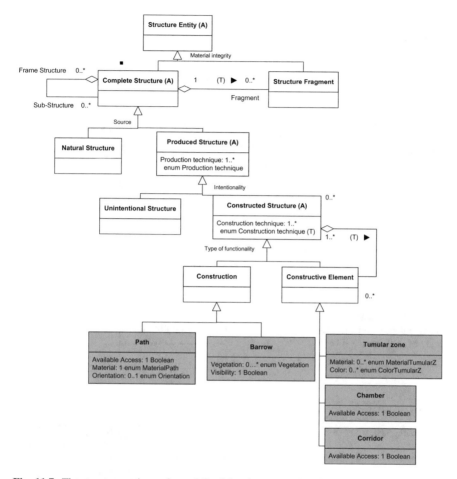

Fig. 11.7 The structure entity package defined for the case study of A Romea

- Corridor: A constructed entity which constitutes a material part of a barrow, to which it contributes structure and/or performs the function of an entrance.

The complete Structure Entity package can be seen in Fig. 11.7.

Last of all, the Valuable Entity package has been defined. This package includes subject sub-packages corresponding to cross-cutting uses of the model resulting from the A Romea extension. In other words, each sub-package within the Valuable Entity package has its own subject matter but with an application which is transversal to the previously defined packages. Figure 11.8 shows the complete composition of the Valuable Entity package. Later, the composition of each sub-package will be described.

The Sample sub-package includes classes which conceptualize representative portions taken from other Valuable Entities represented in the previous packages:

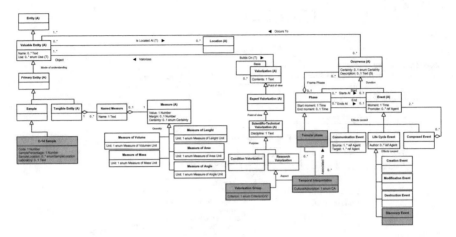

Fig. 11.8 The valuable entity package defined for the case study of A Romea. The classes and associations in orange represent classes and associations which are particular to the CHARM extension created

Fig. 11.9 The sample package defined for the case study of A Romea

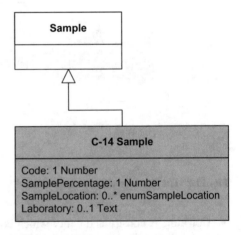

objects, structures and stratigraphic units. It has been necessary to include C-14 Sample in the same extended class for our case study:

- C-14 Sample: A tangible entity corresponding to a fragment of another tangible entity, the properties of which it aims to represent. The fragment is specially picked for a C-14 radiocarbon analysis.

Figure 11.9 shows the composition of classes of the Sample package.

The Measure sub-package includes classes of use in the conceptualization of units of measurement which allow Valuable Entities presented in other packages to be given a dimension. Figure 11.10 shows the composition of classes of the Measure package.

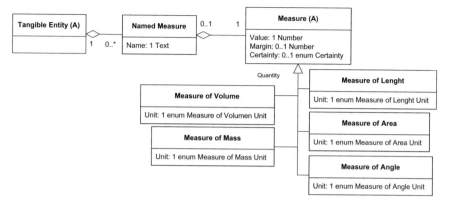

Fig. 11.10 The measure package defined for the case study of A Romea

Fig. 11.11 The location package defined for the case study of A Romea

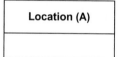

The Location sub-package includes classes of use for the geographic and spatial location of the Valuable Entities which are present in other packages. Figure 11.11 shows the composition of classes of the Location package, which has only one class (Location) present in CHARM and, therefore, maintains its original definition. Due to the fact that the case study of A Romea does not extremely focus its information on spatial location, it has not been necessary to incorporate any extended classes into this package, in spite of the fact that CHARM possesses location sub-classes which offer more detail in this area.

The Occurrence sub-package includes classes of use for the conceptualization of temporal information about the Valuable Entities which are present in other packages. Two classes have been included in the extension for our case study:

- Tumular Phase: A circumstance which takes a relatively long time and corresponds to a stable period of a barrow studied.
- Discovery Event: A circumstance related to the life of an entity which takes a relatively short time and corresponds to the initial moment of a process of learning something that was not known before or of discovering someone or something which was missing or hidden about the entity studied.

Figure 11.12 shows the composition of classes of the Occurrence package.

Lastly, the Valorization sub-package includes classes of use for the conceptualization of valorizations made regarding the entities present in other packages. In this package, the classes *Valorization Group* and *Temporal Interpretation* have been added. The definitions of the extended classes are as follows:

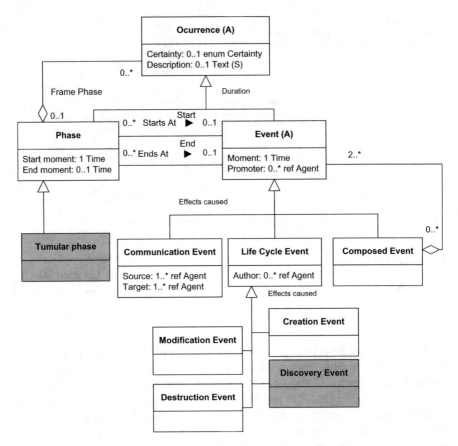

Fig. 11.12 The ocurrence package defined for the case study of A Romea

- Valorization Group: a scientific-technical valorization produced with the purpose of generating new knowledge about the valorized object. The valorized object is a group of stratigraphic units which performs a unit in terms of a methodological, functional or other criterion selected.
- Temporal Interpretation: a scientific-technical valorization produced with the purpose of generating new knowledge about the valorized object. This valorized object could be any Valuable Entity. The new knowledge generated is always related with temporal assignments to the valorized object, interpreting it in temporal terms.

Figure 11.13 shows the composition of classes of the Valorization package.

The packages defined allow us to structure the extension model for A Romea into subject areas with a fixed structure. In addition, the defined framework allows the use of the ConML clusters mechanism in order to identify both the classes with relevant semantics and those which play an auxiliary role in the subject model. The definition of clusters acts as a model view (TypeModelView in the ConML specification) transversally to the packages. Therefore, a cluster may contain classes

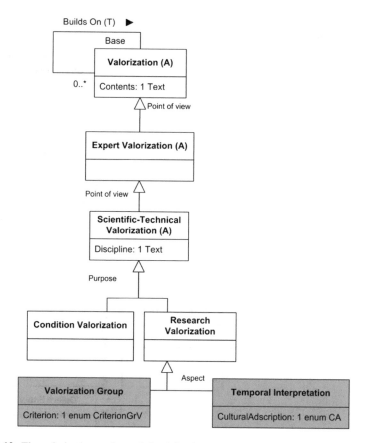

Fig. 11.13 The valorization package defined for the case study of A Romea

belonging to several packages, thus converting this mechanism into the appropriate way of conceptualizing aspects which are transversal to the fixed package structure. Specifically, the model resulting from the A Romea extension consists of 11 clusters, expressed below in ConML notation [129]:

- **Stratum by Deposit–Deposit**: allows us to know which stratigraphic unit of a stratum by deposit type is defined according to which deposit among those found. Figure 11.14 shows the classes belonging to the cluster.
- **Stratum by Object–Object Entity**: allows us to know which stratigraphic unit of a stratum by object type is defined according to which object among those found. Figure 11.15 shows the classes belonging to the cluster.
- **Tangible Entity–Named Measure–Measure**: allows us to know all types of measurements taken from tangible entities. Figure 11.16 shows the classes belonging to the cluster.
- **Tangible Entity–Sample**: allows us to know to which tangible entity the sample taken corresponds. Figure 11.17 shows the classes belonging to the cluster.

Fig. 11.14 Stratum by
deposit–deposit cluster

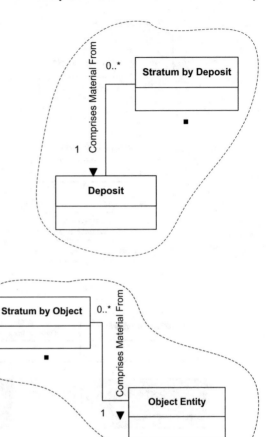

Fig. 11.15 Stratum by
object–object entity cluster

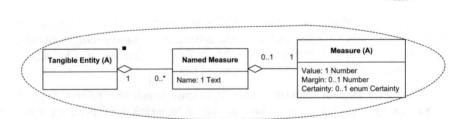

Fig. 11.16 Tangible entity–named measure–measure cluster

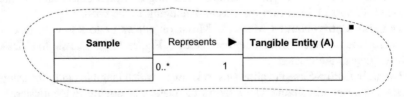

Fig. 11.17 Entity–sample cluster

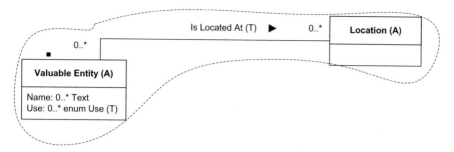

Fig. 11.18 Valuable entity–location cluster

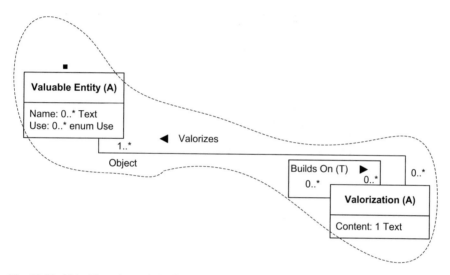

Fig. 11.19 Valuable entity–valorization cluster

- **Valuable Entity–Location:** allows us to know the locations associated to a defined Valuable Entity. Figure 11.18 shows the classes belonging to the cluster.
- **Valuable Entity–Valorization**: allows us to know the valorizations which any agent makes regarding a defined Valuable Entity. Figure 11.19 shows the classes belonging to the cluster.
- **Complete Object–Object Fragment**: allows us to know which fragments correspond to which object among those found. Figure 11.20 shows the classes belonging to the cluster.
- **Complete Structure–Structure Fragment**: allows us to know which fragments correspond to which structure among those found. Figure 11.21 shows the classes belonging to the cluster.
- **Composed Event–Event**: allows us to know which sub-events are included in a given event. Figure 11.22 shows the classes belonging to the cluster.

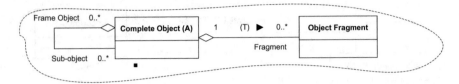

Fig. 11.20 Complete object–object fragment cluster

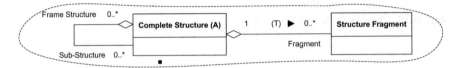

Fig. 11.21 Complete structure–structure fragment cluster

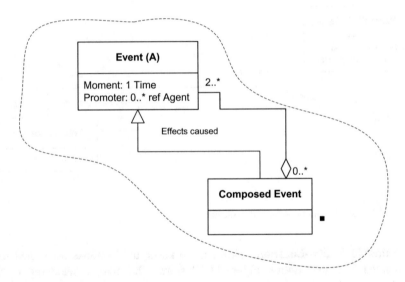

Fig. 11.22 Composed event–event cluster

- **Construction–Constructive Element–Constructed Structure**: allows us to know which constructive elements and other constructions form part of each structure built. Figure 11.23 shows the classes belonging to the cluster.
- **Stratigraphic Unit–Stratigraphic Sequence:** allows us to know to which stratigraphic sequence the defined stratigraphic units belong. Figure 11.24 shows the classes belonging to the cluster.

It should be noted that the main class of the cluster is referenced first in its name. The rest of the classes are participants of the cluster. In this point, it is necessary to remember that the cluster mechanism is used throughout the framework in a

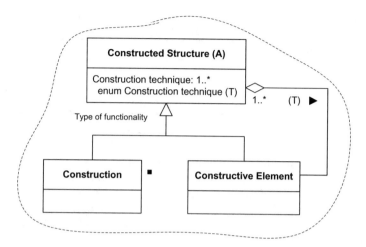

Fig. 11.23 Construction–constructive element–constructed structure cluster

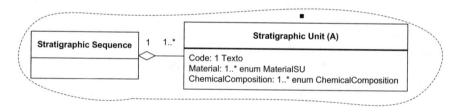

Fig. 11.24 Stratigraphic unit–stratigraphic sequence cluster

hierarchical way. In other words, the cluster is defined at the highest possible level of abstraction, with this characteristic being inherited among the specialized classes which make up the cluster. In addition, it should be highlighted that the cluster mechanism will allow us to then identify the relevant semantics, having this information in the integration model defined for the framework (see Chap. 10).

As part of the application process of the framework proposed for the A Romea case study, this conceptual model has been implemented as a relational database and then instanced in the database itself. The model of instances for A Romea constitutes the implementation of the subject model for our case study. The complete outline of the model implemented can be seen in Chap. 12.

A Romea: Cognitive Processes Model

The proposed framework also has a second perspective to be dealt with, one which is particularly relevant for the design of a software assistance system for the generation of knowledge in archaeology: the cognitive processes which the specialist in

Fig. 11.25 Instances of each
type of inference for the case
study of A Romea in UML
notation

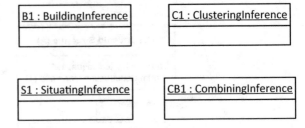

the field carries out in order to ascend in the DIKW hierarchy [7] and, thus, generate new knowledge based on existing archaeological data. In order to provide support for this dimension, a description was given in the Chap. 8 of the metamodel created, which conceptualizes the types of inferences supported by the framework, their sub-classes and the coherence relations which may be present in each discourse and their relations with the most common types of inferences in archaeology. For the specific case of A Romea, the instance for Cognitive Processes has an instance for each type of inference which we wish to assist, as can be seen in Fig. 11.25.

It should be noted that, as was described in the Chap. 8, the types of inference dealt with by the framework are always the same, so in each case of application there will be instances of the four identified types. The existence of instances regarding the coherence relations which are supporting the type of inference will depend on whether the discourse analysis has been carried out for each case of application dealt with and, therefore, this information can be incorporated into the cognitive processes model, providing additional information and information regarding traceability with the discourse of the case of application. In this case, the decision has been taken to maintain only instances for the types of inference due to reasons of simplicity as they are the minimum instances necessary in order to explain in detail the way of working of the proposed framework.

The cognitive processes carried out by archaeologists in order to provide answers to the research questions posed in A Romea (outlined in the previous section "A Romea as a case study: A general overview") are those which must be assisted by the framework, corresponding in our model to the instances of types of inference.

Once we have the particular model of cognitive processes for our case study, with instances for each one of the four types of inference acting as primitives of cognitive processes, the framework can use this information in the integration model (see Chap. 10) and establish assistance in the form of visualization and interaction mechanisms which will be provided in each case. The following section will describe the presentation and interaction mechanisms defined for the case of A Romea.

A Romea: Presentation and Interaction Model

In addition to providing support for cognitive processes and analyzing the subject matter of the case study, the proposed framework should offer a formal representation solution for presentation and interaction in A Romea which allows for software assistance to the analysis of the data from the case study and the generation of knowledge in its context. In order to do this, the proposed solution, which was explained in detail in the Chap. 9 has been applied. This solution proposed a set of RIA patterns organized hierarchically as a reusable repository of solutions. The application of this solution to the specific case study which concerns us here has been carried out by characterizing the problems identified in that chapter and adjusting them to the particular case of A Romea, in such a way that there is a degree of traceability between the abstract challenges presented in that chapter, the specific problems in archaeology and how those problems manifest themselves in A Romea specific case study. In this point, A Romea specifies these problems in the following way:

PROBLEM ARomea_1: As we have seen in the section of this chapter "A Romea as a case study: A general overview", the information obtained from the case study is organized into a single complete dataset which holds information of many different types related with the case study of A Romea, provided by many different researchers, treating this complete dataset as a group. Therefore, a general overview of the information with which we are working is needed, both on a structural level and regarding the relation between that structure and the specific data it holds, so that the different archaeologists can maintain a common idea about the structure of the information in this specific case, what terminology is used (for example, if the term barrow or megalith is employed), etc.

PROBLEM ARomea_2: The complete dataset from A Romea has an abundance of quantitative information, especially regarding the catalogued evidence found during the process of excavation of the barrow, which basically refers to structures and objects. It is necessary, therefore, to have a mechanism for presentation and interaction which enables simple and agile grouping of the evidence found during the excavation of the barrow according to a criterion which the user can vary. In this way, the archaeologists will be able to group together and categorize the evidence present in A Romea in a rapid and agile manner.

PROBLEM ARomea_3: The information present in the complete dataset from A Romea has a great degree of detail in its description, being able to identify, for example, classes of the extension created which group together 8 or 10 attributes in total, due to the mechanism of inheritance in the hierarchies of specialization defined. This situation requires, therefore, the handling of levels of importance throughout the visualization in different interaction units of the information, obtaining, if desired, additional information about a piece of evidence, an event or reference, or a subset of them and maintaining a general view of all of them on the screen.

PROBLEM ARomea_4: The complete dataset from A Romea has an abundance of information about material entities, such as the evidence found during the excavation or the stratigraphic units. All of this information presents cross-cutting characteristics which allow an analysis of the similarities and differences between the data present to be carried out. It is necessary, therefore, to use interface mechanisms which allow the similarities and differences in these subsets of data to be shown.

PROBLEM ARomea_5: The research objectives proposed for the case study of A Romea present a significant temporal component. In fact, its main objective is the tracing of the temporal phases of the barrow of A Romea itself. In this context, it is necessary to present the data reflecting the temporal facet as a main aspect adapted to the case study, paying special attention to the dispersion and vagueness of the temporal data. The geographic aspect is also present in the case study of A Romea, although as it is a limited geographic area, no special needs for presentation and interaction have been identified beyond those already proposed as a general solution for archaeology in the Chap. 9.

PROBLEM ARomea_6: As we are dealing with information extracted as the result of archaeological excavations, the complete dataset from A Romea has an abundance of information organized sequentially or by levels, in this case corresponding to the stratigraphic sequence of the site. As has already been mentioned, the sequential structure of this type of information could go unnoticed without an interface which explicitly takes in this type of data and the relations between its instances. It is necessary, therefore, to use interface mechanisms which allow the data of the case study related to the stratigraphic sequence of the site to be presented in an adapted way.

Having enumerated the problems to be dealt with in the formal representation of the presentation and interaction in our case study, the application of the proposed solution in A Romea has the following structure:

As can be seen in Fig. 11.26, an analogous combination of individual patterns and interaction units to the one presented as a proposed framework for the field of archaeology has been chosen, instancing one interaction unit per type and the individual patterns which correspond to each one, according to the specification and reasons explained in the Chap. 9, with the aim of fully illustrating the proposal. In the case of dealing with other archaeological case studies or focusing on problems of knowledge generation of a different kind, the reader, acting as a practitioner, could define configurations of different patterns, thus providing the framework in general, and the presentation and interaction model in particular, with a great degree of flexibility.

Over the course of the rest of the chapter, and in the later Chap. 12, the question of how these patterns have been adapted in order to assist to the problems of the specific case of A Romea will be dealt with.

A Romea: Integration, Interoperability and Consistency Between Models

Lastly, the case study must be completed with the formal expression of the relations which exist between the subject model, the cognitive processes model and the model of presentation and interaction mechanisms defined and explained above for A Romea.

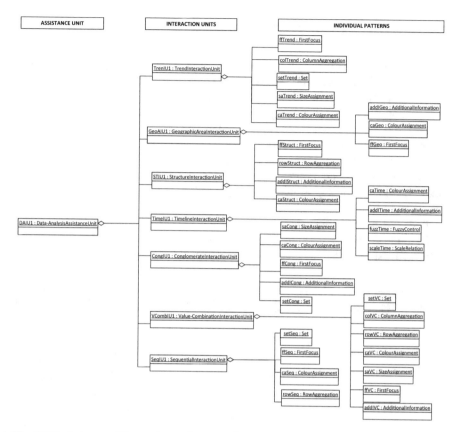

Fig. 11.26 Instances of interaction and presentation patterns for the case study of A Romea in UML notation

In order to achieve this, we shall apply the integration metamodel built with this in mind and described in the Chap. 10. It should be remembered at this point that, according to this integration metamodel, it is necessary for the three models involved (the Subject Model, the Processes Model and the Interaction Model) to be instanced, in order to, then, express, via instances of the integration metamodel (which acts as a pivot), the relations between them.

Over the course of this chapter, we explain in detail how the information about the formal relations between the three parts of the framework are obtained, how this information is reflected in the instances of the integration metamodel proposed for the framework and how these instances allow the framework to carry out the software assistance to the generation of knowledge in archaeological A Romea case which we seek, offering adapted software presentation and interaction patterns.

Obtaining Information on Integration

As has been described in Part III, the framework presented acquires the structure and content of the information upon which knowledge will be generated thanks to the subject model. From this subject model, the integration metamodel obtains information about which packages and clusters are defined, what are the main classes and characteristics for each cluster and in which packages the main class of each cluster is present (see the relations between instances of *ClusterMainClassDescription* and *PackageDescription* in Fig. 11.27). In addition, the framework has a cognitive processes model which allows us to define the most relevant cognitive processes in archaeology, which we have instanced for A Romea (see Fig. 11.25).

It should be noted that the packages which interest us thematically in order to assist to the generation of knowledge are those which have the attribute *Aux* with a value of False, that is to say, the packages which are semantically main in the case

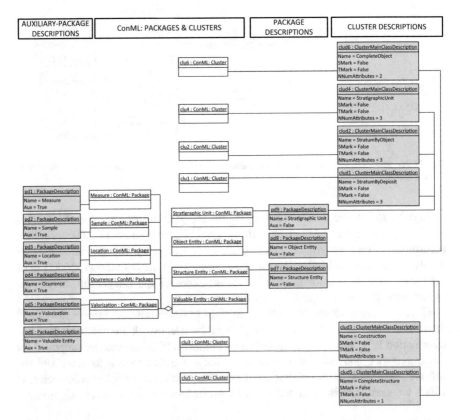

Fig. 11.27 Information extracted from the subject model of A Romea. The figure shows all the packages defined and the clusters with direct relation to them (6 clusters of the 11 defined in the complete model for A Romea)

study with which we are concerned. In this case, there are three: *Stratigraphic Unit*, *Object Entity* and *Structure Entity*. The clusters which have a direct relation with them, as seen in Fig. 11.27, are those which we shall also deal with on the level of interoperability. For each one of the clusters with a direct relation with the packages of the model, four types of cognitive processes are evaluated, which we have defined for archaeology, instancing one object for each type.

With this information, the case study of A Romea provides us with the following matrix of possibilities:

It should be noted that each cell formed by the cross between cluster-cognitive process evaluated from this matrix would correspond to an instance of the *InteractionOption* class in the integration metamodel. In each cell of Table 11.1, the interaction units which the system can offer for assistance to the generation of knowledge in the field according to each case are defined. For the definition of one or another interaction unit within each cell, the universal guidelines defined in Chap. 10 on Interoperability have been applied, along with those for application to the archaeological particularities. It should be remembered that the universal guidelines are:

Table 11.1 Matrix of instance possibilities for package defined versus cognitive processes assisted, and the corresponding interaction units defined in each case

		Building inference	Clustering inference	Combining inference	Situating inference
Package	Related cluster				
Stratigraphic unit	Stratigraphic unit	-Structure IU	-Value-Combination IU -Conglomerate IU	-Value-Combination IU - Trend IU	-Geographic IU
	Stratum By deposit	-Structure IU	-Value-Combination IU -Conglomerate IU	-Value-Combination IU - Trend IU	-Geographic IU
	Stratum By object	-Structure IU	-Value-Combination IU -Conglomerate IU -Sequential IU	-Value-Combination IU -Trend IU	-Geographic IU
Object entity	Complete object	-Structure IU	-Value-Combination IU -Conglomerate IU	-Value-Combination IU -Trend IU	-Geographic IU
Structure entity	Complete structure	-Structure IU	-Value-Combination IU -Conglomerate IU	-Value-Combination IU	-Geographic IU
	Construction	-Structure IU	-Value-Combination IU -Conglomerate IU	-Value-Combination IU -Trend IU	-Geographic IU -Timeline IU

- Each cluster defined and each inference of a Building Inference type can determine an interaction option which refers to a Structure IU interaction unit. This interaction unit enables assistance to problems, such as Problem 1 defined in A Romea, related with the necessity of exploring the internal structure of the dataset of the case study
- Each cluster defined and each inference of a Clustering Inference type can determine two interaction options:

  (1) An interaction unit of a Conglomerate IU type, which allows the data to be grouped by the values associated to instances belonging to the cluster's main class. This interaction unit allows us to assist to problems such as Problem 2 defined for A Romea, which are related with the necessity for dynamism in the visualization of datasets.
  (2) An interaction unit of a Value-Combination IU type, which allows the data to be grouped by the values associated to attributes and to classes belonging to auxiliary packages which are associated to the cluster's main class. This interaction unit allows us to assist to problems such as Problem 3 defined for A Romea, which are related with the necessity for dealing with levels of importance throughout the visualization, obtaining, if desired, additional information on subsets of evidence, events or other instances, whilst maintaining on the screen the general view of the cluster being dealt with.

- Each cluster defined and each inference of a Combining Inference type can determine two interaction options:

  (1) An interaction unit of a Value-Combination IU type, which allows the values associated to attributes and to classes belonging to auxiliary packages which are associated to the cluster's main class to be explored. This interaction unit allows us to assist to problems such as Problem 3 defined for A Romea, which are related with the necessity for dealing with levels of importance throughout the visualization, obtaining, if desired, additional information on a piece of evidence, an event or a reference or a subset, whilst maintaining on the screen the general view of the cluster being dealt with.
  (2) An interaction unit of a Trend IU type in cases in which the value of the NNumAttributes attribute of the CusterMainClassDescription class is higher than 2. This implies that the main class of the cluster dealt with presents various numerical, enumerated or Boolean attributes, which are potentially susceptible to varying considerably from one version of the dataset to another. This interaction unit allows us to assist to problems such as Problem 4 defined for A Romea, which are related with the necessity for carrying out analyses of similarities or differences between the data at different stages of the research

In archaeological cases, it is necessary to apply specific guidelines in order to decide the most appropriate interaction unit(s), identifying:

- If the main class of the package concerns stratigraphic information and the cognitive process to be evaluated regards Clustering, a Sequential IU unit is defined, which allows Problem 6 defined for A Romea to be assisted, which is related with the specific visualization of sequential information, of a stratigraphic nature, and its specific analysis.
- Each cluster defined and each inference of a Situating Inference type can determine two interaction options:

  (1) An interaction unit of a Geographic Area IU type in cases in which the main class of the cluster presents relations with the auxiliary package Location. This interaction unit enables us to assist to problems such as Problem 5 defined for A Romea, which is related with the need for visualizing geographic locations of the entities studied.

  (2) An interaction unit of a Timeline IU type in cases in which the main class of the cluster presents relations with the auxiliary package *Occurrence*. This interaction unit enables us to assist to problems such as Problem 5 defined for A Romea, which is related with the need for visualizing information with a strong temporal aspect regarding the entities studied.

In summary, the information obtained thanks to the subject model of A Romea, along with the evaluation of the cognitive processes defined on the level of clusters applying these rules, allows the framework to offer the appropriate interaction units in order to provide assistance to archaeologists in these cognitive processes.

It should be noted that the instances of the cognitive processes evaluated, along with the interaction options and interaction units defined in Table 11.1, are objects in the integration model for the case of A Romea. Due to the complexity of the resulting model, the decision was taken to illustrate the integration model by selecting the instances corresponding to one non-auxiliary package of the three of which A Romea consists: Structure Entity. The complete integration model presents similar models of instances of integration for each one of the remaining non-auxiliary packages of the case study (Stratigraphic Unit and Object Entity). Figure 11.28 shows the model of instances of the integration metamodel for the Structure Entity package.

Implementing Interoperability

Up to this time, we have had the complete scenario of the case of A Romea at our disposal, described by the team involved and modelled according to the meta-models of the framework for the three perspectives which concern us: subject matter, cognitive processes and presentation and interaction mechanisms. In addition, we have the instance of the integration metamodel, which defines interoperability between the three perspectives for the case study, as well as which guidelines

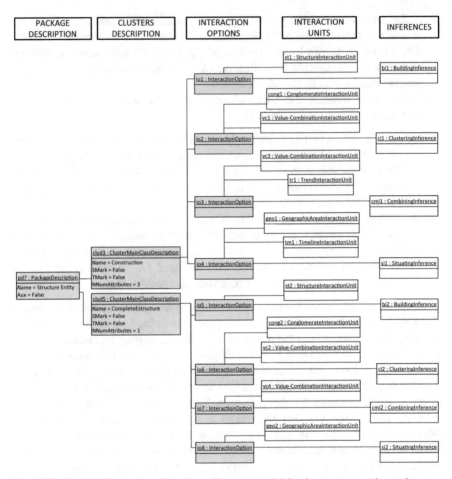

Fig. 11.28 Model of instances of the integration metamodel for the structure entity package

(both universal and those specific to the case being dealt with) will be applied for software assistance to the generation of knowledge in the case in question.

Both the models of the three perspectives and the integration model which acts as a pivot have been subjected to a first validation, being instanced for a specific case (not A Romea) with archaeological data [175, 179]. Its design and validation follow the interoperability framework defined by Pastor [211], which establishes the conditions which need to be fulfilled for an interoperability solution for models in order to guarantee interoperability on syntactic, semantic and technical levels. All of the validation was carried out using tools from the Eclipse EMF suite (*Eclipse Modeling Framework*) [87], and can be seen in detail in [175].

It should be taken into account that, during the validation in [175], the application scenario was narrated by archaeologists, who had no prior knowledge of the structure of the metamodels involved or of the integration metamodel. In the same way as in this first validation, the archaeologists also narrated the case study of A Romea, its objectives, the structure of the information dealt with and the new knowledge which was generated in a free way. This separation of activities between the conception of the integration metamodel and the application scenario allows us to establish if the integration metamodel is capable of correctly expressing the relations detailed, avoiding the creation of an ad hoc integration metamodel for the proposed scenario.

As a result, and basing ourselves on the initial validation carried out, we can affirm that the proposed integration metamodel is capable of expressing the relations between the different aspects or perspectives which are involved in a software-assisted knowledge generation system (in terms of subject matter, cognitive processes and presentation and interaction mechanisms) to a high level of abstraction. This permits the materialization of software assistance by way of adapted visualizations, showing acceptable behavior in terms of structural cohesion and semantics in the context of EMF. For more information on the EMF implementation carried out, see [87]. Furthermore, the proposed metamodel has been capable of expressing the real archaeological case study.

Conclusions

This chapter presents the validation of the framework proposed and guides the reader for putting it into practice in a real archaeological case study. During the course of this chapter, the various solutions provided in terms of the subject model, cognitive processes and software presentation and interaction mechanisms for A Romea have been applied. The application of these different solutions have given rise to specific software models which materialize the proposal presented in order to assist archaeologists in the generation of knowledge based on the large datasets and the information of the case study in question. In addition, this chapter shows that the integration of the different modeling perspectives in a common conceptual representation (the integration metamodel), is possible, implemented through open-source technology, such as [87], among others previously detailed.

It can, therefore, be stated that the framework fulfils the desired objectives as far as the capacity for reflecting what kind of software-assisted knowledge generation in the form of information visualization mechanisms are to be offered is concerned. Applying the proposed framework to any other case study in a similar context would give rise to specific models of application of similar characteristics to those presented throughout this analytical validation, allowing the characterization of the structure of information which the specific case presents, identifying which of the defined processes we wish to assist (it is not necessary to assist them all) and which application of the pattern metamodel is to be carried out in each case. This lends the framework a great deal of flexibility in terms of specifying what type of software assistance based on information visualization mechanisms we desire but adjusting it to the needs of archaeological research which we have identified throughout the course of the research. The possibilities of applying, supporting and using the framework presented in other contexts and/or fields, as well as providing other types of software assistance to the generation of knowledge will all be dealt with in detail in the Chap. 13.

Chapter 12
Empirical Validation

Introduction

The previous chapter presented a formal validation of the software models which make up the framework presented in this book, applied to the specific case study of A Romea. This validation demonstrates the capacity of the framework to provide the kind of software-assisted knowledge generation for archaeological datasets, which we aim to offer by way of interaction and presentation patterns adapted to the structure of archaeological information and to the cognitive processes performed on this information during the knowledge generation process.

However, we were of the opinion that it was necessary to go one step further and create a prototype of the system resulting from the implementation of these framework models for the specific case of A Romea. This prototype has enabled us to carry out an empirical validation with archaeologists of the software assistance provided by the framework, showing it real benefits to analyze archaeological large datasets. It should be remembered at this point that, as we explained in the presentation of the case study, A Romea is a real archaeological case of knowledge generation, which was carried out years before this research began. This enabled the software assistance offered by the framework to be evaluated by comparing it to the method employed for knowledge generation in the original case.

This chapter describes the process of implementation and prototyping of the framework presented, along with the empirical validation carried out in collaboration with archaeological specialists using the afore-mentioned prototype. This empirical validation complements the previous analytical validation, in an attempt to verify the original hypothesis of this research. Furthermore, it allows us to guide the readers on how design, implement and apply the framework to build a real software system based on the framework proposed. In addition, the empirical validation determines the degree of software assistance achieved and detect problems and points to improve on in the future.

© Springer International Publishing AG 2018
P. Martin-Rodilla, *Digging into Software Knowledge Generation in Cultural Heritage*, Modeling and Optimization in Science and Technologies 11,
https://doi.org/10.1007/978-3-319-69188-6_12

Prototyping Process and Prototype Characteristics

The software models resulting from the framework's application for the specific case of A Romea provide us with a full structure in terms of the subject model, the assisted cognitive processes and the well-defined software presentation and inter-action mechanisms. This structure has served as a reference point for a software system prototyping the framework.

We have developed a prototype which implements the full framework using iOS technology [124], version 8.4 SDK (Software Development Kit) [125] for devel-opment in Apple iPad devices. The memory which the device temporarily dedicates to the application enables the subject model's implemented structure to be saved in the relational database thanks to the CoreData library [126] (which is integrated into the SDK). Our main motivation when choosing iOS as the basis for the prototype was born out of the prior expertise of the author of this book in mobile technolo-gies. Furthermore, having the prototype in tablet format sped up the empirical validation with end users as it enabled us to carry out the empirical validation in different archaeological institutions without the need to install the framework on different devices and gave us greater logistical advantages. However, it should be noted that this prototype is only one of the several possibilities of types of appli-cations that we can built using the framework presented, being possible their application to other technological contexts that we need to analyze large archaeo-logical datasets. For example, in web environments for collaborative work or in desktop application for a more detailed data laboratory work in a specific area of archaeological data. Besides, although the technological environments used for the prototype development will change over time, the framework structure, method-ology and guidelines of application will still be applicable due to the abstraction level of definition employed.

Therefore, the prototype built consists of:

(1) A part of the subject model, in which the thematic structure of the case of A Romea has been implemented as a relational database (included in a reference section at the end of this chapter).
(2) Software controllers, which, instancing the defined interoperability metamodel, enable us to interpret and decide what kind of software assistance is most suitable according to the cognitive process selected by the user.
(3) The defined interaction and presentation mechanisms. Finally, these interaction and presentation mechanisms have been implemented as interfaces, following the mock-ups showed in previous chapters.

We shall go on to describe how, during the design phase of the empirical validation, we identified the need for implementing not only the A Romea case study, which has been fully developed throughout this book, in the prototype, but also another archaeological case study. This decision was taken in order to prevent the A Romea case study from interfering in the results of the empirical validation.

Therefore, a second case study, named "Forno dos Mouros" was chosen and implemented in the same way that has already been described for A Romea.

The resulting prototype allows the archaeologist to select which case study he/she wishes to analyze (see in next pages Fig. 12.4). Then, he/she selects which of the four types of cognitive processes instanced in the cognitive processes model he/she wants to perform. The archaeologist must also select which package is to be studied from among those defined in the extension of the case study. The structural information of the package and the data of the case study itself can be found in the relational database implemented for each case study.

After having selected these two options, the framework instances the interoperability metamodel and offers the user the available interaction units, according to the instances which have been created from the defined interaction and presentation metamodel (all the possibilities for A Romea can be seen in the Chap. 11). Due to the need to develop another full case study, in addition to the characteristics of the selected cases themselves, the prototype implements a subset of the matrix of possibilities of each case study.

As far as the specific case of A Romea is concerned, the previous Table 12.1 shows the matrix of possibilities for assistance offered. The interaction units in bold are those which are implemented in the prototype. In order to illustrate this more clearly, Fig. 12.1 shows the instance of the presentation and interaction metamodel defined for the A Romea case study, with the classes implemented by the prototype shaded in blue.

It should be noted that four of the six types of interaction units defined are implemented. The two exceptions are Trend IU (due to the fact that the case studies do not present much need to observe trends, as there is only one dataset for each case) and Geographic IU (due to the fact that, in both cases, the knowledge generation process being assisted does not present specific questions of a geographic nature, as both cases are archaeological sites with a clear geographic location without patterns of comparison with other geographic references). The case study of "Forno dos Mouros" follows a similar structure, implementing Structure IU in order to assist in Building inferences, Value-Combination IU for Clustering and Combining inferences and Timeline IU for Situating inferences. In figures from Figs. 12.2, 12.3, 12.4, 12.5, 12.6, 12.7, 12.8, 12.9 and 12.10 some screenshots of the implemented prototype are shown.

This prototype enables us to test the implementation of the defined framework and to identify possible problems in its effective use. As the readers can see, using the framework we can go step by step defining the aspects of our information that we are going to analyze and assist, and what kind of interface mechanisms we want to use for doing it. Next section act as a reference guide with the real subject model implemented in the prototype as a relation database for the A Romea case study, in order to complete the prototyping information provide to our readers.

Table 12.1 Matrix of possibilities for assistance offered, bold interaction units are offered through the framework prototype

Package	Cluster	Building inference	Clustering inference	Combining inference	Situating inference
Stratigraphic unit	Stratigraphic unit	– Structure IU	– Value-combination IU – Conglomerate IU	– Value-combination IU – Trend IU	– Geographic IU
	Stratum by deposit	– Structure IU	– Value-combination IU – Conglomerate IU	– Value-combination IU – Trend IU	– Geographic IU
	Stratum by object	– structure IU	– Value-combination IU – Conglomerate IU – Sequential IU	– Value-combination IU – Trend IU	– Geographic IU
Object entity	Complete object	– Structure IU	– Value-combination IU – Conglomerate IU	– Value-combination IU – Trend IU	– Geographic IU
Structure entity	Complete structure	– Structure IU	– Value-combination IU – Conglomerate IU	– Value-combination IU	– Geographic IU
	Construction	– Structure IU	– Value-combination IU – Conglomerate IU	– Value-combination IU – Trend IU	– Geographic IU – Timeline IU

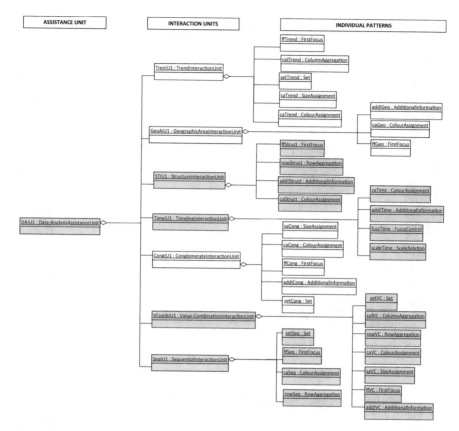

Fig. 12.1 An instance of the framework's presentation and interaction metamodel applied to the case study of A Romea. The elements implemented by the prototype are shaded in blue

A Romea Prototype Database Implementation

This section presents the structure and implementation of A Romea subject model performed, and includes:

- The complete implementation as a relational database of CHARM model.
- The implementation of the extension created to support the particularities of the A Romea case study.
- The conceptualization and implementation in a relational database format of the mechanisms to use Cluster and Packages explained in Chap. 11.

Note that all structures—CHARM and the extension created for the case study— support multilingual definition. Thus, the tables which includes the suffix "_Xlat" corresponds to language specifications.

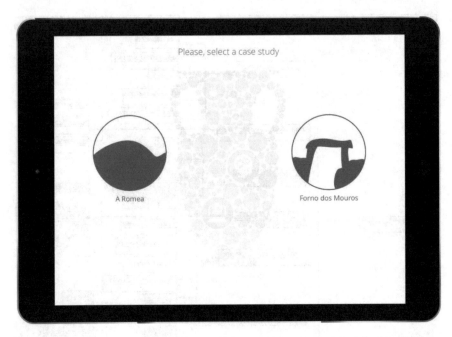

Fig. 12.2 The home screen of the implemented framework's prototype: the selection of the case study

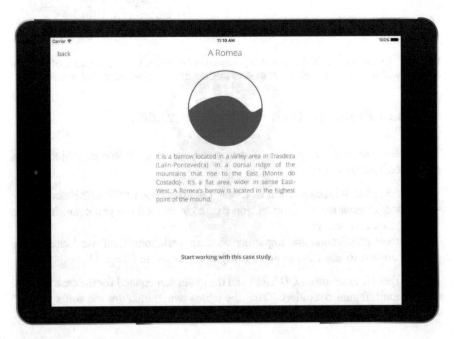

Fig. 12.3 Screen showing a summary of the selected case study

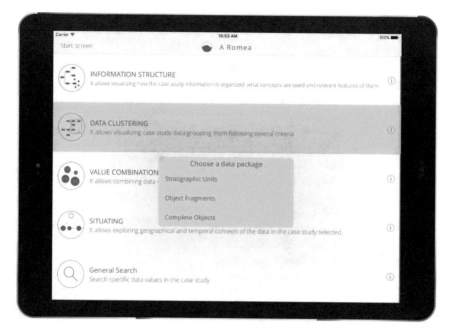

Fig. 12.4 The selection screen for the cognitive process to be assisted

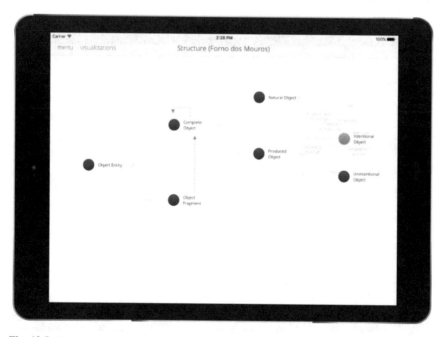

Fig. 12.5 A screenshot showing the final implementation of the Structure IU interaction unit for the Forno dos Mouros case study (Objects package)

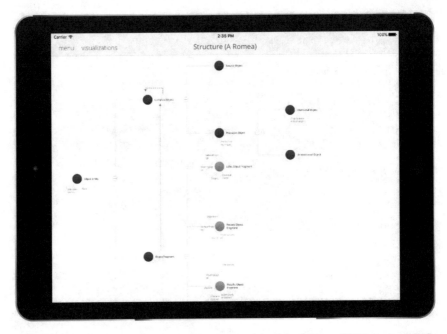

Fig. 12.6 A screenshot showing the final implementation of the Structure IU interaction unit for the A Romea case study (Objects package)

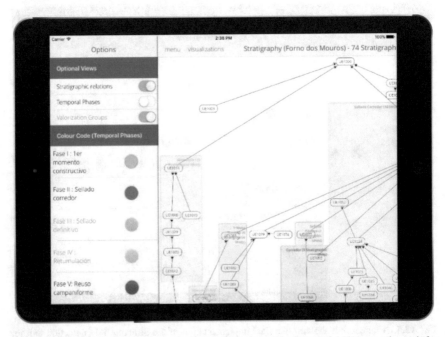

Fig. 12.7 A screenshot showing the final implementation of the Sequential IU interaction unit for the Forno dos Mouros case study

Fig. 12.8 A screenshot showing the final implementation of the Value-Combination IU interaction unit for the A Romea case study

Fig. 12.9 A screenshot showing the final implementation of the Value-Combination IU interaction unit for the A Romea case study showing the configuration options

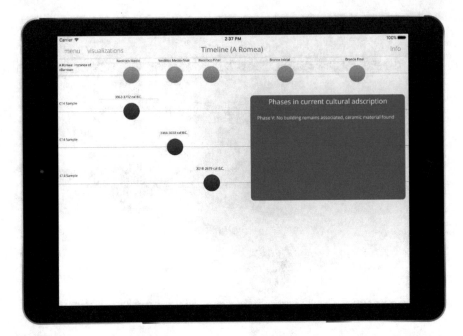

Fig. 12.10 A screenshot showing the final implementation of the Timeline IU interaction unit for the A Romea case study

The Subject Model

Tables and relations specification

1. **Table: _Languages**

   Columns

Name	Data type	Size
Id	Long integer	4
Code	Text	2

   Number of items: 2
   Relations

Relation	Source table	Destination table	Relation features
1 to N	_Languages	_Languages_Xlat	Forced; Cascade delete
1 to N	_Languages	BaseTypes_Xlat	Forced
1 to N	_Languages	Classes_Xlat	Forced
1 to N	_Languages	EnumeratedItems_Xlat	Forced

(continued)

(continued)

Relation	Source table	Destination table	Relation features
1 to N	_Languages	EnumeratedTypes_Xlat	Forced
1 to N	_Languages	Generalizations_Xlat	Forced
1 to N	_Languages	Packages_Xlat	Forced
1 to N	_Languages	SemiAssociations_Xlat	Forced

2. Table: _Languages_Xlat

Columns

Name	Data type	Size
Id	Long integer	4
Owner	Long integer	4
Language	Long integer	4
TheName	Text	32

Number of items: 4
Relations

Relation	Source table	Destination table	Relation features
1 to N	_Languages_Xlat	_Languages	Forced; Cascade delete

3. Table: Associations

Columns

Name	Data type	Size
Id	Long integer	4
PrimarySemiAssociation	Long integer	4
SecondarySemiAssociation	Long integer	4

Number of items: 124
Relations

Relation	Source table	Destination table	Relation features
1 to N	Associations	Links	Forced
1 to N	Associations	SemiAssociations	Forced

4. Table: Attributes

Columns

Name	Data type	Size
Id	Long integer	4
OwnerClass	Long integer	4
RedefinedOriginal	Long integer	4
MinCard	Long integer	4
MaxCard	Long integer	4
IsSorted	Boolean Yes/No	1
BaseType	Long integer	4
EnumeratedType	Long integer	4
IsTemporal	Boolean Yes/No	1
IsSubjective	Boolean Yes/No	1

Number of items: 80
Relations

Relation	Source table	Destination table	Relation features
1 to N	Attributes	Attributes_Xlat	Forced; Cascade delete
1 to N	Attributes	Values	Forced
1 to N	Attributes	BaseTypes	Forced
1 to N	Attributes	Classes	Forced
1 to N	Attributes	EnumeratedTypes	Forced

5. Table: Attributes_Xlat

Columns

Name	Data type	Size
Id	Long integer	4
Owner	Long integer	4
Language	Long integer	4
TheName	Text	64
Definition	Memo	–

Number of items: 160
Relations

Relation	Source table	Destination table	Relation features
1 to N	Attributes_Xlat	_Languages	Forced
1 to N	Attributes_Xlat	Attributes	Forced; Cascade delete

6. Table: BaseTypes

Columns

Name	Data type	Size
Id	Long integer	4

Number of items: 5
Relations

Relation	Source table	Destination table	Relation features
1 to N	BaseTypes	Attributes	Forced
1 to N	BaseTypes	BaseTypes_Xlat	Forced; Cascade delete

7. Table: BaseTypes_Xlat

Columns

Name	Data type	Size
Id	Long integer	4
Owner	Long integer	4
Language	Long integer	4
TheName	Text	64
Definition	Memo	–

Number of items: 10
Relations

Relation	Source table	Destination table	Relation features
1 to N	BaseTypes_Xlat	_Languages	Forced
1 to N	BaseTypes_Xlat	BaseTypes	Forced; Cascade delete

8. Table: Classes

Columns

Name	Data type	Size
Id	Long integer	4
IsCHARM	Boolean Yes/No	1
IsAbstract	Boolean Yes/No	1
IsTemporalAspect	Boolean Yes/No	1

(continued)

(continued)

Name	Data type	Size
IsSubjectiveAspect	Boolean Yes/No	1
DominantGeneralization	Long integer	4
Package	Long integer	4
ClusterRoot	Boolean Yes/No	1

Number of items: 183
Relations

Relation	Source table	Destination table	Relation features
1 to N	Classes	Attributes	Forced
1 to N	Classes	Classes_Xlat	Forced; Cascade delete
1 to 1	Classes	Clusters	Forced
1 to N	Classes	Generalizations	Forced
1 to N	Classes	Objects	Forced
1 to N	Classes	ParticipantClassesInClusters	Forced; Cascade delete
1 to N	Classes	SemiAssociations	Forced
1 to N	Classes	SpecializedClassesOfGeneralizations	Forced
1 to N	Classes	Generalizations	Forced
1 to N	Classes	Packages	Forced
1 to N		ClassesInPackages	Not forced

9. Table: Classes_Xlat

Columns

Name	Data type	Size
Id	Long integer	4
Owner	Long integer	4
Language	Long integer	4
TheName	Text	64
Definition	Memo	–
Comments	Memo	–

Number of items: 366
Relations

Relation	Source table	Destination table	Relation features
1 to N	Classes_Xlat	_Languages	Forced
1 to N	Classes_Xlat	Classes	Forced; Cascade delete

10. Table: ClassesInPackages

Columns

Name	Data type	Size
Package	Long integer	4
ParticipantClass	Long integer	4

Number of items: 100
Relations

Relation	Source table	Destination table	Relation features
1 to N	ClassesInPackages	Packages	Not forced
1 to N	ClassesInPackages	Classes	Not forced

11. Table: Clusters

Columns

Name	Data type	Size
Id	Long integer	4
IsCHARM	Boolean Yes/No	1
MainClass	Long integer	4

Number of items: 11
Relations

Relation	Source table	Destination table	Relation features
1 to 1	Clusters	Classes	Forced
1 to N	Clusters	ParticipantClassesInClusters	Forced; Cascade delete

12. Table: EnumeratedItems

Columns

Name	Data type	Size
Id	Long integer	4
Type	Long integer	4
SuperItem	Long integer	4

Number of items: 245
Relations

Relation	Source table	Destination table	Relation features
1 to N	EnumeratedItems	EnumeratedItems_Xlat	Forced; Cascade delete
1 to N	EnumeratedItems	EnumeratedItems	Forced

13. Table: EnumeratedItems_Xlat

Columns

Name	Data type	Size
Id	Long integer	4
Owner	Long integer	4
Language	Long integer	4
TheName	Text	64
Definition	Memo	–

Number of items: 490
Relations

Relation	Source table	Destination table	Relation features
1 to N	EnumeratedItems_Xlat	_Languages	Forced
1 to N	EnumeratedItems_Xlat	EnumeratedItems	Forced; Cascade delete

14. EnumeratedTypes

Columns

Name	Data type	Size
Id	Long integer	4
IsCHARM	Boolean Yes/No	1
Package	Long integer	4

Number of items: 37
Relations

Relation	Source table	Destination table	Relation features
1 to N	EnumeratedTypes	Attributes	Forced
1 to N	EnumeratedTypes	EnumeratedItems	Forced
1 to N	EnumeratedTypes	EnumeratedTypes_Xlat	Forced; Cascade delete
1 to N	EnumeratedTypes	Packages	Forced

15. EnumeratedTypes_Xlat

Columns

Name	Data type	Size
Id	Long integer	4
Owner	Long integer	4
Language	Long integer	4
TheName	Text	64
Definition	Memo	–
Comments	Memo	–

Number of items: 74
Relations

Relation	Source table	Destination table	Relation features
1 to N	EnumeratedTypes_Xlat	Attributes	Forced
1 to N	EnumeratedTypes_Xlat	EnumeratedTypes	Forced; Cascade delete

16. Table: Generalizations

Columns

Name	Data type	Size
Id	Long integer	4
GeneralizedClass	Long integer	4

Number of items: 76
Relations

Relation	Source table	Destination table	Relation features
1 to N	Generalizations	Classes	Forced
1 to N	Generalizations	Generalizations_Xlat	Forced; Cascade delete
1 to N	Generalizations	SpecializedClassesOfGeneralizations	Forced; Cascade delete
1 to N	Generalizations	Classes	Forced

17. Table: Generalizations_Xlat

Columns

Name	Data type	Size
Id	Long integer	4
Owner	Long integer	4
Language	Long integer	4
Discriminant	Text	64

Number of items: 152
Relations

Relation	Source table	Destination table	Relation features
1 to N	Generalizations_Xlat	_Languages	Forced
1 to N	Generalizations_Xlat	Generalizations	Forced; Cascade delete

18. Table: Packages

Columns

Name	Data type	Size
Id	Long integer	4
IsCHARM	Boolean Yes/No	1

Number of items: 9
Relations

Relation	Source table	Destination table	Relation features
1 to N	Packages	Classes	Forced
1 to N	Packages	EnumeratedTypes	Forced
1 to N	Packages	ClassesInPackages	Not forced
1 to N	Packages	Packages_Xlat	Forced; Cascade delete

19. Table: Packages_Xlat

Columns

Name	Data type	Size
Id	Long integer	4
Owner	Long integer	4

(continued)

(continued)

Name	Data type	Size
Language	Long integer	4
TheName	Text	64
Description	Memo	–
Comments	Memo	–

Number of items: 18
Relations

Relation	Source table	Destination table	Relation features
1 to N	Packages_Xlat	_Languages	Forced
1 to N	Packages_Xlat	Packages	Forced; Cascade delete

20. Table: ParticipantClassesInClusters

Columns

Name	Data type	Size
Cluster	Long integer	4
ParticipantClass	Long integer	4

Number of items: 13
Relations

Relation	Source table	Destination table	Relation features
1 to N	ParticipantClassesInClusters	Clusters	Forced; Cascade delete

21. Table: SemiAssociations

Columns

Name	Data type	Size
Id	Long integer	4
OwnerClass	Long integer	4
RedefinedOriginal	Long integer	4
MinCard	Long Integer	4
MaxCard	Long integer	4
IsSorted	Boolean Yes/No	1

(continued)

(continued)

Name	Data type	Size
IsWhole	Boolean Yes/No	1
IsStrong	Boolean Yes/No	1
IsTemporal	Boolean Yes/No	1
IsSubjective	Boolean Yes/No	1

Number of items: 248
Relations

Relation	Source table	Destination table	Relation features
1 to N	SemiAssociations	Classes	Forced
1 to N	SemiAssociations	Associations	Forced
1 to N	SemiAssociations	Associations	Forced
1 to N	SemiAssociations	SemiAssociations_Xlat	Forced; Cascade delete

22. Table: SemiAssociations_Xlat

Columns

Name	Data type	Size
Id	Long integer	4
Owner	Long integer	4
Language	Long integer	4
TheName	Text	64
Role	Text	64
Definition	Memo	–

Number of items: 496
Relations

Relation	Source table	Destination table	Relation features
1 to N	SemiAssociations_Xlat	_Languages	Forced
1 to N	SemiAssociations_Xlat	SemiAssociations	Forced; Cascade delete

23. Table: SpecializedClassesOfGeneralizations

Columns

Name	Data type	Size
Generalization	Long integer	4
SpecializedClass	Long integer	4

Number of items: 184
Relations

Relation	Source table	Destination table	Relation features
1 to N	SpecializedClassesOfGeneralizations	Classes	Forced
1 to N	SpecializedClassesOfGeneralizations	Generalizations	Forced

24. Table: Objects

Columns

Name	Data type	Size
Id	Long integer	4
Identifier	Text	64
TypeClass	Long integer	4
Description	Text	255

Number of items: 3461
Relations

Relation	Source table	Destination table	Relation features
1 to N	Objects	Classes	Forced
1 to N	Objects	Links	Forced
1 to N	Objects	Links	Forced
1 to N	Objects	Values	Forced

25. Table: Links

Columns

Name	Data type	Size
Id	Long integer	4
TypeAssociation	Long integer	64

(continued)

(continued)

Name	Data type	Size
PrimaryObject	Long integer	4
SecondaryObject	Long integer	4
PhaseSelector	Long integer	4
PerspectiveSelector	Long integer	4
Description	Short text	255

Number of items: 4050
Relations

Relation	Source table	Destination table	Relation features
1 to N	Links	Associations	Forced
1 to N	Links	Objects	Forced
1 to N	Links	Objects	Forced

26. Table: Values

Columns

Name	Data type	Size
Id	Long integer	4
OwnerObject	Long integer	4
TypeAttribute	Long integer	4
Contents	Short text	255
PhaseSelector	Long integer	4
PerspectiveSelector	Long integer	4

Number of items: 11,601
Relations

Relation	Source table	Destination table	Relation features
1 to N	Values	Attributes	Forced
1 to N	Values	Objects	Forced

Validation Methodology

The prototype implemented, in addition to as a framework's application guide, has served as a tool for the proposed empirical validation, which is especially interesting in order to ensure that our solution it's improving the knowledge generation process in some archaeological data analysis processes. The following section describes the full design of the validation process and how it was carried out employing this prototype, constituting a methodological guide for any reader who wants to test rigorously, with real archaeological specialists, its own applications built following the provided framework. The experience during several works carried out in designing, evaluating and monitoring empirical strategies in the field of Software Engineering in Cultural Heritage fields [171, 173, 177] led us to choose Wohlin's framework [284] as a reference guide for the empirical validation which concerns us here.

The Validation Process

According to Wohlin's reference framework [284], we shall define below each of the stages of the experimentation process followed for the empirical validation of the software models used in order to assist in the generation of knowledge. These models have been generated throughout this research and are referred to as the framework proposed (software models affecting three dimensions: the subject model, the cognitive processes model and the software interaction and presentation model. Furthermore, it has an interoperability model which describes the relations between them).

We design an evaluation process with real users (archaeologists) from public and private heritage institutions in Spain, which is the country of institutional affiliation of the author. The empirical validation of the framework created is described below, following the Wohlin's steps.

Definition of the Empirical Study Range and Objectives— Scoping

The objective of the empirical validation in Wohlin's terms [284] can be defined as follows:

> To compare the framework proposed with the traditional method employed in the process of generating knowledge from raw data in archaeology, with the aim of evaluating the quality of the software assistance provided to the process of knowledge generation from the point of view of archaeologists in the context of public and private institutions in Spain.

Planning

Context

The validation empirical study is set in a specific, though broad, context as the subjects are all archaeologists and related professionals belonging to public and private institutions on a national scale. The context, therefore, is considered to be the professional environment of the subjects. The objects used are created ad hoc for the empirical study, although they use real data from the field of archaeology.

The Formulation of Hypotheses

There are many characteristics which can be compared between the traditional method of generating knowledge from raw data and the proposed framework. These include aspects of usability, the study of processes carried out during the analysis itself, the study of decision-making and the role played by the use of one specific method or another, the degree of comprehension and handling, as well as aspects relating to integrity, flexibility and portability of both methods and their related technology.

However, because the ultimate aim of the framework is to provide software assistance to the generation of knowledge in archaeology, we have only selected characteristics of interest for the discipline. More specifically, we deal with **accuracy**, **efficiency** and **productivity**, which are achieved with both methods when carrying out data analysis tasks for the generation of knowledge. Furthermore, we believe it is relevant, given the assistance nature of the framework, to study aspects of the user's (in our case the archaeologist) **satisfaction**. Finally, the **quality of the knowledge generated** with both methods must be considered, with the aim of fully evaluating the assistance function of the proposed framework. The research questions which aim to deal with each of these characteristics and the hypotheses which arise from them in this empirical validation are as follows:

- **RQ1**: Does the framework affect **accuracy** in tasks relating to knowledge generation in archaeology? We define accuracy as a "quantitative measure of the magnitude of error" [136]. We shall measure accuracy as the percentage of correct answers that the subjects give once the defined knowledge generation tasks have been carried out. Therefore, the null hypothesis shall be:

  H_{01}: The accuracy in tasks relating to the generation of knowledge in archaeology using the framework is similar to the accuracy obtained when carrying out the same tasks with the traditional method of data analysis.

- **RQ2**: Does the framework affect efficiency in tasks relating to knowledge generation in archaeology? We define efficiency as "the ability to produce a result with a minimum of extraneous or redundant effort" [135]. We shall measure efficiency as the response time employed by the subjects in carrying out the defined knowledge generation tasks. Therefore, the null hypothesis shall be:

H_{02}: The efficiency in tasks relating to the generation of knowledge in archaeology using the framework is similar to the efficiency obtained when carrying out the same tasks with the traditional method of data analysis.

- **RQ3**: Does the framework affect the **productivity** of archaeologists when generating knowledge in archaeology? We define productivity as "the ratio of work product to work effort" [135]. We shall measure productivity as the ratio between the accuracy achieved and the response time employed in carrying out the indicated data analysis tasks. Therefore, the null hypothesis shall be:

H_{03}: The productivity in tasks relating to the generation of knowledge in archaeology using the framework is similar to the productivity obtained when carrying out the same tasks with the traditional method of data analysis.

- **RQ4**: Does the framework affect the archaeologist's **satisfaction** when generating knowledge in archaeology? We define satisfaction as the degree of "positive attitudes towards the use of the product" [137]. We shall measure satisfaction as the degree of ease of use expressed by the subjects when carrying out the tasks, via the use of a Likert questionnaire with a 5-point scale. Therefore, the null hypothesis shall be:

H_{04}: The satisfaction expressed by subjects when carrying out tasks relating to the generation of knowledge in archaeology using the framework is similar to the satisfaction they express when carrying out the same tasks with the traditional method of data analysis.

- **RQ5**: Does the framework affect the **quality of the knowledge generated** by the archaeologists? We define quality as "the degree to which a product meets specified and implicit requirements [137] when it is used in specific conditions" [138]. As we are interested in specifically measuring the quality of the generated knowledge, we must know whether this knowledge meets the needs of the archaeologists. This generated knowledge generally takes the form of text (archaeological reports, monographs or similar documents). Thus, a written report in archaeology shall be evaluated as an end product. We shall measure this internal quality of the generated knowledge by analyzing the correctness and satisfaction of the archaeologists when evaluating the written report produced, using the data analysis methods being compared. Correctness shall be measured as the number of errors reported by the archaeologists when evaluating the end product by evaluating a report made during the analysis of each of the data analysis methods. Satisfaction shall be measured as the degree of general perception expressed by the archaeologists when evaluating the end product, using a Likert questionnaire [163] with a 5-point scale. Therefore, the null hypothesis shall be:

H_{05}: The degree of correctness and satisfaction expressed by subjects evaluating the generated knowledge as a product (a written report) of the data analysis process in archaeology using the framework is similar to the degree of correctness and satisfaction they express when carrying out the same evaluation with the traditional method of data analysis.

The Selection of Variables

Response Variables

The selected response variables emerge from the five characteristics listed above as being especially relevant when comparing the traditional method of data analysis and the proposed framework. We can define these variables as follows:

- **Accuracy**: the subject's magnitude of error when carrying out data analysis tasks using both methods is measured. The metric selected shall be the percentage of correct answers that the subjects give once they have carried out the defined knowledge generation tasks. The tasks are divided into several items so it is possible to obtain the percentage of accuracy by measuring the percentage of items which have been successfully completed in all the tasks. The total accuracy obtained can be aggregated in two ways: Firstly, by only taking into account the percentage corresponding to the tasks which have been carried out, paying correct attention to all their sub-items (in other words, the tasks which have been carried out fully and correctly). Secondly, accuracy can also be aggregated by calculating the average accuracy obtained for each task, thus avoiding the interference of aspects relating to the type of task at that moment, although this data may be of interest in future analyses.
- **Efficiency**: measuring the subject's ability to carry out tasks relating to the generation of knowledge from raw data, thus providing answers to relevant questions in the generation of new knowledge in the proposed archaeological problems by using as few resources as possible. In this case, the selected metric is the response time the subject takes to carry out the tasks. The total accuracy obtained is aggregated by calculating the average accuracy obtained for each tasks, thus avoiding interference from aspects relating to the type of task at that moment, although this data may be of interest in future analyses.
- **Productivity**: the ratio between work and resources achieved by the subject using both methods to carry out data analysis tasks is measured. Therefore, the metric will be that ratio.
- **Satisfaction**: the degree of ease of use for both methods is measured. As we are working in assistance contexts, we believe it is relevant to evaluate this variable, as assistance can be obstructed due to an inappropriate degree of ease of use. The metric used will be the ease of use expressed according to a 5-point Likert scale. The instrument employed is a questionnaire based on Moody's framework [190], which evaluates satisfaction based on three concepts: Perceived Usefulness (PU), Perceived Ease of Use (PEOU) and Intention to Use (ITU). In accordance with Moody's framework, a questionnaire regarding ease of use for

each treatment with 22 sentences for evaluation, of which eight evaluate perceived usefulness (PU), nine evaluate ease of use (PEOU) and five evaluate intention to use (ITU). Each questionnaire on treatment refers integrally to all the tasks carried out with the same treatment, following the same structure as the rest of the response variables.

- **The quality of the generated knowledge**: the degree of quality which the generated knowledge presents using both methods is measured. In order to do so, we have selected two relevant sub-aspects in this generated knowledge: (1) the degree of correctness and (2) the degree of satisfaction of the subjects as far as the new knowledge generated is concerned.

The correctness metric consists of the number of errors reported by the specialists in the field, analyzing the generated knowledge by way of the analysis of the written report, according to the method used to produce it. The majority of knowledge generated in archaeology is found in written texts, as several published studies have already pointed out (see Chap. 4). Therefore, it is important to use this format in order to evaluate the degree of correctness of the knowledge which is generated. After finishing the data analysis tasks with the selected treatment, the archaeologists produced a written report of between 300 and 500 words, in which they summarized the case study analyzed and drew conclusions about it. Later, they evaluated the written text of another subject, reporting errors following a script provided which tackles four types of errors:

- Errors, in the opinion of the specialist, in references found in the written report regarding the raw data of the case study analyzed.
- Errors, in the opinion of the specialist, in phrases alluding to the processes which the author carried out in order to draw conclusions from the data. For example, the use of verbs such as to compare, to classify, to identify, to categorize, to differentiate, to be the cause or consequence of, etc. or exemplifications and/or generalizations.
- Errors, in the opinion of the specialist, due to redundancies in the written report regarding an aspect which has already been dealt with in another part of the same report, such as the repetition of arguments, references to raw data and examples.
- Errors, in the opinion of the specialist, in phrases alluding to the conclusions which the author comes to regarding the case study.

The satisfaction metric consists of the degree of positive attitude expressed by the archaeologists when evaluating this knowledge by way of the written reports produced by other subjects. This has been measured using a questionnaire with sentences regarding the written reports and evaluated via a 5-point Likert scale.

Factor

For each of the variables identified, we shall apply the factor (variable controlled by the responsible of each empirical study, which is applied on different levels in order to discover its impact on the model). We shall call our factor "the data analysis method". It should be noted that this term possesses specific semantics at the heart of this validation: the data analysis method consists of a set of tasks which are performed in order to examine raw data with the aim of extracting conclusions and thus generate new knowledge. Generally, these conclusions will then support the decision-making process in the field in question and will verify or refute existing models or theories within that field. Therefore, we are not dealing with tasks relating to data extraction (which are commonly related to the categorization of data) but with tasks which focus on the inferences which emerge from the data. The "data analysis method" is made up of two levels:

M_1: The control level. This is the traditional method of data analysis employed to generate knowledge from raw data in archaeology, based on the direct observation of data obtained from the archaeological case being studied. The data is generally organized in a table format and basic software with limited graphic capacity, such as spreadsheets, is used. In this case, we shall use Excel [185], as it is one of the most widespread tools and is commonly used in the institutions to which the subjects in the validation process belong.

M_2: The treatment level. This is the data analysis method to generate knowledge from raw data in archaeology in which the framework is used to perform the analysis. In this method, the data of the archaeological case being studied is described using a CHARM extension [128] and is selected in the form of subsets by the archaeologist. Depending on the subset selected, the framework presents this data organized into interaction patterns, according to the cognitive process being assisted. In order to increase the legibility of the statistical results of this validation, we shall use the acronym **SAKG** (Software Assisted Knowledge Generation) to refer to the proposed framework.

Table 12.2 shows a summary of the research and hypotheses questions dealt with, the variables and the metrics which will be used in order to define them:

Blocking Variables

The archaeological problem (e.g. the archaeological case study that the archaeologist is analyzing with the prototype: A Romea and Forno dos Mouros) being dealt with in the validation has been detected as a blocking variable and shall be called P from this point on. In order to prevent the archaeological problem from affecting the results of the validation, its value has been balanced, thus blocking the possible effect. In order to do this, P takes two values, P1 and P2. Therefore, the subjects will

Table 12.2 Research questions dealt with in the validation, the defined variables and their corresponding metrics

Research question	Hypothesis	Response variables	Metric
RQ1	H_{01}	Accuracy	Percentage of correct answers
RQ2	H_{02}	Efficiency	Response time
RQ3	H_{03}	Productivity	Accuracy/Efficiency
RQ4	H_{04}	Satisfaction	Perceived usefulness (PU), Perceived ease of use (PEOU), Intention to use (ITU)
RQ5	H_{05}	Quality of generated knowledge: correctness and satisfaction	Number of errors (Correctness) Likert scale (Satisfaction with the generated knowledge)

carry out the data analysis tasks on two different archaeological case studies. Another advantage of using two problems is that the threat of the learning effect is avoided, as what is learnt with the problem of the first treatment is not applied in the second treatment. In next section entitled "Design Principles of the Validation", more details are given about how the P variable has been balanced.

The Selection of Subjects

The subjects are all archaeologists and professional working in archaeology although they could have a different academic background, mainly from heritage backgrounds such as History, Archaeology, the History of Art and Architecture. An open call for participation was made via an e-mail list of archaeological professionals in Spain or of Spanish nationality. In turn, these professionals were encouraged to share this call with other colleagues. The validation process was carried out with 16 volunteers belonging to 7 public (such as the University of Santiago de Compostela, the Institute of Heritage Sciences and the University of Minho) and private (different archaeological companies, the Campo Lameiro Archaeological Park) institutions. The volunteers were selected randomly from among all those who expressed an interest in collaborating in the validation process. Later, the implications of the size and characteristics of the sample taken will be dealt with along with the discussion of the results. In order to characterize the sample in a better way, the subjects completed a demographic questionnaire before starting the process of validation, which they did individually. The demographic distribution of the subjects selected is described below.

General Demographic Data

In this section, we shall examine the demographic characteristics of the sample selected to carry out the empirical validation.

As far as gender is concerned, Fig. 12.11 shows that 56% of the subjects were male and 44% were female. This distribution reflects the proportion by national gender in the archaeology. For example, the INE's (Spanish Statistical Office) 2011 report on University teaching in Spain [132] established that, of 170 university teachers in the area of "Archaeology", 73 were women, thus reflecting a female percentage of 42%, which is similar to the distribution of our sample. Another complementary study was carried out on Human Resources in Science and Technology [131] and examined the system of Science in Spain, analyzing its different areas. The latter report dates from 2009 and shows that 14.70% of doctors work in areas of the Humanities. Of this 14.70, 6.56% were women, representing a relative percentage of 44%, similar to our own sample. We believe, therefore, that this balanced and representative distribution as far as gender is concerned allows us to interpret the data obtained without offering views differentiated by gender.

As far as the subjects' age is concerned, we considered it necessary to carry out the validation with a heterogeneous group, in order to avoid any bias in the sample. Thus, Fig. 12.12 shows that the oldest group, with 38% of the subjects, corresponds to subjects older than 37 years of age. On the other hand, there is significant representation of the other two groups involved, with subjects ranging from 33 to 45 years of age. The lowest percentage, although it is still representative, corresponds to the age range of 26–32 years of age, which generally corresponds to professionals with less work experience. The age ranges indicated have been defined according to the age ranges commonly used in research, in which the first

Fig. 12.11 Percentages of the sample according to distribution by gender

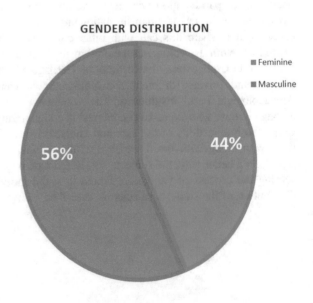

ranges generally consist of staff in training and the older ranges to research staff with a greater level of experience. It should be noted that age is not an indicator of the level of training of the participating subject in all cases. Therefore, this aspect has been dealt with as a variable separated from age. However, the ranges defined are useful given that, in most cases, they also indicate the stage of the subject's research career.

Finally, we carried out a profile study regarding the subjects' level of studies. It should be noted that the level of studies of any subject in the sample will be high compared to a random sample of the general population, due to the characteristics of the end users at whom the study is aimed. A professional in archaeology will generally have a high level of studies so we do not believe that this determines any of the variables being measured in the validation. However, we believe it is interesting to characterize the sample according to the level of studies in order to

Fig. 12.12 Percentages of the sample according to age distribution

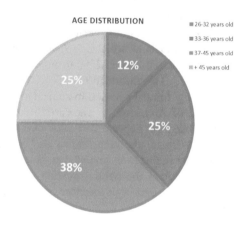

Fig. 12.13 Percentages of the sample corresponding to level of education

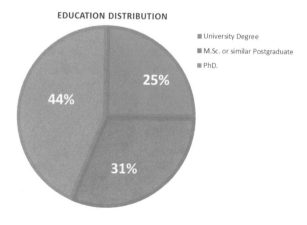

illustrate that, given a random sample of our end users, almost half of them are doctors (see Fig. 12.13). This allows us to gain an idea of what type of people generate knowledge in archaeology and the importance of dealing with reasoning processes, not only defined processes, when it comes to building software assistance in this field.

Therefore, along general lines, we have a sample which is balanced as far as gender is concerned, well-distributed in terms of age and polarized when it comes to the level of education. This last aspect is due to the characteristics of our typical end users: professionals in archaeology who generate knowledge in the field.

Disciplinary Profile

Firstly, we wanted to characterize the sample according to the professional sector to which the subjects belong. In a similar way to the level of education, this variable is illustrated in Fig. 12.14 and has only been used in order to give a more detailed idea of the type of professionals of which the sample consists. Therefore, 56% of the subjects work in the public sector, 25% in private companies and the remaining 19% are self-employed. This information allows us to gain an idea of the enormous importance of publicly funded research in archaeology in Spain. An in-depth study of the specific field of Archaeology was carried out in 2011 [207], studying the proliferation and subsequent debacle after 2007 of the private sector connected with heritage research in Spain, the area known as "commercial archaeology". This study showed that the majority of knowledge generation produced in archaeology in Spain took place within the public research sector, with the support of small companies or individuals forming a structure based on KIBS (Knowledge Intensive Business Services) [206].

Fig. 12.14 Percentages of the sample corresponding to the subjects' professional sector

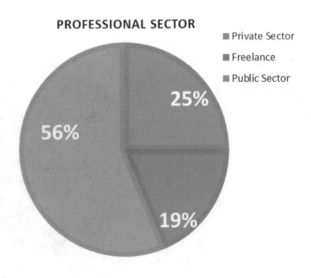

In the following section, we shall characterize the sample according to the sub-disciplines represented. As has been explained before, our main user profiles are archaeologists, but inside this group we can find little differences between their archaeological activity or academic initial background. Taking into account the fact that the objective of this empirical validation is the confirmation of several hypotheses regarding the method of data analysis in two separate case studies, we considered it was necessary to have a heterogeneous sample of most representative situations, albeit one which allowed each of the subjects to have expertise which is in some way related with the subject matter of both case studies. Thus, Fig. 12.15 shows that 44% of the subjects are archaeologists, due to the fact that the case studies selected fundamentally deal with archaeological data. It must be pointed out that, within this 44%, there are archaeologists specialized in different sub-disciplines and chrono-cultural periods, including experts in Metallurgy, Archaeological Science and other areas such as rock art and Prehistoric, Roman and/or Medieval Archaeology, as an example of the internal diversity of this subset. Other groups of experts included architects, restorers, museum managers, art historians, communicators and educators in archaeological matters. Furthermore, the sample has been diversified with experts in the field of geographical information

Fig. 12.15 Percentages of the sample corresponding to disciplinary profile

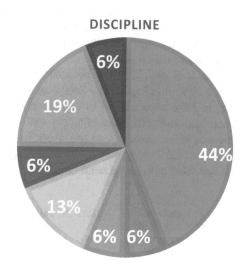

DISCIPLINE

■ Archaeology

■ Preservation And Restoration

■ Heritage Arquitecture

■ Geographical Information Systems applied to Archaeology

■ Art History and Arts

■ Archaeological Divulgation (Museums, Educational programs)

■ Heritage Management

systems applied to archaeology. These experts are able to offer an interesting perspective to the validation as they handle more advanced knowledge generation tools than mere spreadsheets.

Position Profile

Finally, this section aims to offer a more detailed perspective of the sample by characterizing specific aspects of the subjects as far as the knowledge generation in archaeology carried out over the course of their careers is concerned. This was achieved by asking them questions about how much experience they had in handling, managing, documenting and/or researching archaeological data, the primary source for the generation of knowledge in the area.

The ranges shown have been defined according to intuitive intervals reflecting the degree of experience of the subject: someone with less than 6 years of experience is considered to be a professional with a low level of experience, increasing in intervals of 4 or 5 years of experience up to 20 years, at which stage it is considered that the individual has acquired professional maturity. Figure 12.16 shows the distribution of the sample by intervals of years of experience. It should be noted that our sample is quite heterogeneous, although subjects with more than 14 years of experience are predominant. We believe this confers robustness upon the sample when it comes to evaluating the data analysis methods used to carry out knowledge generation processes.

However, it is possible for a professional in archaeology to have spent many years handling archaeological raw data, even working in research, but for their main functions to have been limited to extraction or characterization, with knowledge generation and in-depth archaeological and heritage interpretation tasks being performed by other members of the team. In order to avoid this situation, each subject was asked individually what percentage of his/her working time corresponded specifically to the generation of new knowledge in the field and implied

Fig. 12.16 Percentages of the sample responding to the question "How many years of experience do you have in handling, managing, documenting and/or researching archaeological data?

YEARS OF EXPERIENCE IN ARCHAEOLOGICAL
KNOWLEDGE GENERATION

19%   18%

19%

44%

■ 2-6 years

■ 7- 13 years

■ 14-19 years

■ More than 19 years

Fig. 12.17 Percentage of the sample responding to the following question: What percentage of your career would you say has been dedicated to knowledge generation tasks in archaeology?

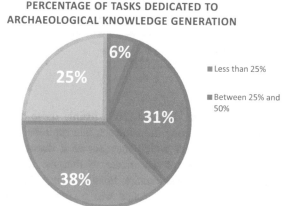

PERCENTAGE OF TASKS DEDICATED TO
ARCHAEOLOGICAL KNOWLEDGE GENERATION

- Less than 25%
- Between 25% and 50%

the use of data analysis methods such as those being evaluated in the empirical validation. Figure 12.17 shows the distribution of answers to this question and how (although the distribution is relatively heterogeneous) the majority of subjects have mainly worked throughout their career on the generation of knowledge. We believe, therefore, that the sample is robust enough in this aspect to be able to evaluate data analysis methods during the processes of knowledge generation.

Design Principles of the Validation

In [205], a thorough review is carried out of the alternatives of validation design in software engineering following the terminology employed by Juristo & Moreno [147] in order to name different design alternatives. Following these studies, it can be observed that the most appropriate alternative for validations involving two treatments (such as that which concerns us) is the **paired design blocked by experimental objects**, in this case, the variable P, which represents the archaeological problem whose data is analyzed. This design presents the following advantages: (1) It maximizes the number of subjects in the validation, as it does not divide the size of the sample in two; (2) It limits the dependence of the problem selected, as we have two problems (P1 and P2); (3) The learning effect [205] between the two treatments is avoided, given that the subjects apply both treatments in different contexts (the defined problems P1 and P2).

Therefore, we selected the paired design blocking our variable P (Problem) for the validation. The subjects we divided into two groups (G1 and G2). Both groups used the traditional data method (M_1) in the first session (S1) and the method based on our SAKG framework proposed (M_2) in the second session (S2). Table 12.3 shows the design applied:

It should be pointed out that, although the design chosen avoids the majority of the threats to validity previously specified, some threats, which are intrinsic to this

Table 12.3 A summary of the design of the validation

		P1	P2
Session 1	M_1	G1	G2
Session 2	M_2	G2	G1

design, have been detected [205], mainly those due to the context of the sessions (noise, fatigue among the subjects, interruptions, etc.), which may vary in S1 and S2. Another aspect which may act as a threat to validity is the profile of the subjects. An attempt was made to mitigate this aspect by carrying out initial measuring via a questionnaire. Finally, it should be highlighted that the assignation of the subjects to the groups was carried out randomly, maintaining the same number of subjects in both groups in order to keep them balanced.

Instrumentation

The objects employed in the empirical validation were:

- An initial questionnaire regarding the subjects' profile (gender, age, level of education, discipline and years of professional experience with archaeological information).
- A statement with the required data analysis tasks, which were different for problem 1 (P1) and for problem 2 (P2).
- Excel files with the real data employed in both problems (P1 and P2).
- Working prototypes of the framework with the real data of both problems (P1 and P2).
- Two evaluative questionnaires for both methods (M1 and M2).
- A document of guidelines and a satisfaction questionnaire for the written reports.

All of the documents used in the validation can be consulted in next sections in the order in which they are listed above.

Archaeological Problems

The paired design model blocked by experimental objects selected for this validation requires two separate problems (P1 and P2) to be defined in order to carry it out. The defined archaeological problems (P1 and P2) are described below.

Problem 1: A Romea

Problem 1 consists of carrying out an analysis of the data available regarding the historical and archaeological evidence found during the excavation of the archaeological site known as "A Romea" [169]. The case study is located in the excavation and prospection of a metallurgical zone in which objects of different materials, compositions and temporal and functional origins were found. In addition, parts of a barrow were documented via the discovery of structures, attributing temporal phases to the barrow: when it was built, when it was used, when it was abandoned, etc. The data analysis should determine the functional attributions of the archaeological site and the objects found, as well as the temporal phases associated to it based on those objects and structures. In order to do this, there is data available regarding its material composition, morphology, decoration and other aspects of interest for archaeologists concerning the objects and structures found. Furthermore, radiocarbon dating was available for some of the structures, along with geographic information for each documented element.

Problem 2: Forno dos Mouros
Problem 2 consists of carrying out an analysis of data available about historical and archaeological evidence found during the excavation of the site known as "Forno dos Mouros" [6, 159]. The case study is located in the excavation of a megalithic structure used for burials, in the style of a dolmen, in which objects and structures are documented. In addition, the dolmen has the remains of paintings inside. The data analysis must determine the temporal attributions of the objects and structures found. In order to do this, data is available regarding the material composition, morphology, decoration and other aspects of interest for archaeologists concerning the objects, structures and paintings found. Furthermore, geographic information for each documented element is available.

Both problems respond to two archaeological case studies.

Data Analysis Tasks

As the factor being applied is the "method of data analysis", it is necessary to define a series of data analysis tasks to be performed by the subjects, enabling the five selected variables to be evaluated. Taking into account the archaeological field in which the validation in set, and previous studies regarding what cognitive processes are carried out in this field in order to generate knowledge [172], certain tasks relating to these cognitive processes have been defined. Each task emphasizes the performance of a specific cognitive process. Thus, the definition is similar for both problems, only varying in terms of the structure and content of the information being analyzed in each problem (the answer will vary according to the problem dealt with due to this fact). The statements for the tasks are as follows:

- TASK A: This task concerns processes of the combination of values in order to find out the distribution and characteristics of the materials found in any archaeological study.

  - Statement PROBLEM 1: Indicate and note down the total number of ceramic fragments, lithics and/or other materials which have been found at the site.

Later, calculate again the totals of the ceramic fragments, lithics and/or other materials but, this time, also according to the method of extraction used.

– Statement PROBLEM 2: Indicate and note down the total number of ceramic fragments, lithics or other materials which have been found at the site. Later, calculate again the totals of the ceramic fragments, lithics or other materials but, this time, also according to whether the fragments are decorated or not.

- TASK B: This task concerns grouping processes.

  – Statement PROBLEM 1: Indicate and note down the percentage of average fragmentation presented in the ceramic objects extracted mechanically.
  – Statement PROBLEM 2: Indicate and note down the percentage of average fragmentation presented in the ceramic objects extracted manually.

- TASK C: This task concerns processes of contextual situation in order to discover the temporal characteristics of the objects and/or material structures found in the archaeological studies.

  – Statement PROBLEM 1: Indicate and note down the different chronological attributes which you believe have been associated with the objects found in the site.
  – Statement PROBLEM 2: Indicate and note down the different chronological attributes which have been associated with the ceramic fragments found in the site. Then, reason about the cultural attributes of the complete ceramic pieces: Are you able to associate a chrono-cultural attribute to each ceramic object? Indicate the ones associated to pieces PZ01 and PZ06.

- TASK D: This task concerns processes of the combination of values in order to discover functional aspects of the materials and structures found in the archaeological studies.

  – Statement PROBLEM 1: Indicate how many stratigraphic units have been defined and according to which criteria the groups of these stratigraphic units have been created. Then, indicate which units make up the "Barrow Chamber" group and the "Alteration Path" group.
  – Statement PROBLEM 2: Indicate how many stratigraphic units have been defined and according to which criteria the groups of these stratigraphic units have been created. Then, indicate which units make up the "First megalithic zone" group and the "Chamber access: pit" group.

- TASK E: This task concerns processes of the analysis of the internal structure of the data.

  – Statement PROBLEM 1: Indicate which attributes (relevant characteristics) of the objects found have been documented.
  – Statement PROBLEM 2: Indicate which attributes (relevant characteristics) of the stratigraphic units found have been documented.

- TASK F: This task is related to processes of contextual situation in order to discover the temporal characteristics of the objects and/or material structures found in the archaeological studies.

  - Statement PROBLEM 1: Indicate and note down which temporal intervals (chronological assignments) have been attributed to the A Romea barrow in a global manner, paying attention to all the information you are given regarding the stratigraphy of the materials found.
  - Statement PROBLEM 2: Indicate and note down here which temporal intervals (chronological assignments) have been attributed to the Forno dos Mouros barrow in a global manner, paying attention to all the information you are given regarding the stratigraphy of the materials found.

- TASK G: Answer the following question: **What archaeological conclusions do you draw from the case study in question?** by writing a brief text of approximately 300–500 words.
- TASK H: Point out, using the template created for this purpose, the errors which, in your opinion, can be observed in the written report which you have been given to evaluate.

The written reports produced are reflections of the knowledge generated as a result of the data analysis. Tasks G and H will serve, therefore, to evaluate these written reports at a later time in terms of the quality of the knowledge generated via the metrics defined for this purpose (correctness and satisfaction). By neither referring to the structure of the information nor to the specific content of each problem dealt with, problems P1 and P2 both have the same statement.

The Development Dynamics of the Empirical Validation

As far as the lines of action in carrying out the study are concerned, the subjects (our volunteer archaeologists) were given some brief instructions about the dynamics and they were allowed to ask some initial questions. No previous training of the subjects was considered necessary as far as the methods to be used were concerned: the traditional method of data analysis was well known by all the subjects, whereas the method of analysis using the framework was not known beforehand. This difference is assumed within the validation in order to find out how intuitive the framework is, although we are aware that some prior training in the latter method could affect (hopefully in a positive way) the results offered by the subjects with this method. However, our interest in finding out how intuitive the proposed method is, along with recommendations made in empirical studies in software engineering [284], led us to take the decision not to offer any prior training in either of the two methods, assuming the difference in familiarity of the subject with both methods and the difference in the degree of expertise between the two. The dynamics of the empirical validation were divided into two sessions (S1 and S2), as can be seen in Table 12.3. Each session consisted of specific phases:

1. In the first phase, the subjects filled in the demographic questionnaire, which was the same for everyone. Then, the subjects were divided into two groups, chosen at random. This phase was only carried out in the first session of the validation.
2. In the second phase, the subjects performed tasks A, B, C, D, E and F applying the method (or treatment) assigned in each case to the problem selected, depending on whether it was session S1 or session S2, according to the design shown in Table 12.3. Therefore, this phase was carried out in both sessions. Once the tasks were completed, each user filled in the corresponding satisfaction questionnaire, independently of the group to which they belonged.
3. In the third phase, each subject wrote a report of less than 500 words about the case study dealt with in the session. It should be noted that, depending on the group to which the subject belonged, he/she had performed the tasks following one method and on one problem in each session. He/she, based on the data analysis performed, had to answer the question: What archaeological conclusions do you draw about the case study dealt with? This phase was also carried out in both sessions.
4. In the fourth phase, each subject had to evaluate a report written by a different subject. The possibility of one subject evaluating his/her own report was avoided. This task corresponded to task H and was carried out at the end of the second session. The subjects were able to evaluate reports written after using either method of data analysis and regarding either of the two problems. The full dynamics of the validation can be seen in Fig. 12.18.

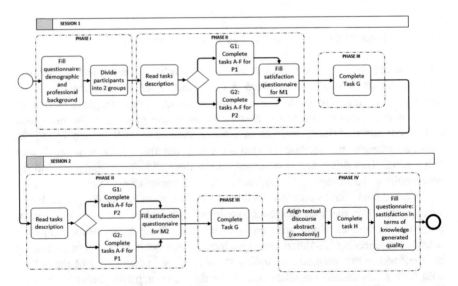

Fig. 12.18 The development dynamics of the empirical validation, indicating the phases corresponding to each session. The rectangles represent tasks, whereas the diamonds indicate the beginning of tasks performed in parallel

An Evaluation of Validity Aspects

Due to the need for empirical validation of the framework presented in this book, it was considered necessary to carry out an in-depth analysis of all the threats to its validity which could present themselves, in order to guarantee the validity of the results produced. Due to this fact, on this occasion, the threats identified in the literature in the field of Software Engineering (Wohlin [284], based on Cook & Campbell [54]), have been analyzed one by one in an attempt to identify those which may influence the validation presented and to describe how they have been mitigated (see from Tables 12.4, 12.5, 12.6 and 12.7).

Operation

Preparation: The subjects in the empirical validation did not know the case studies being used nor were they familiar with the data. They were given minimal information about method 2 in the cases in which they needed to use that method, though they were not offered any prior training or informed of the hypotheses being dealt with in the study. By offering themselves as volunteers for the validation, they gave their consent to these aspects. The necessary materials referred to in the earlier section entitled "Instrumentation" were prepared in advance.

Execution: The subjects attended two sessions of between 60 and 90 min in length.

Validation of the data: No invalid data was detected. This can be attributed to the individuality and supervision of the process of executing the study.

Analysis and Interpretation

Other studies carried out in software engineering with a paired design blocked by experimental objects [72, 147, 205] were taken into account for the analysis and interpretation of the results. These studies generally use the general linear model with repeated measures (thereby maximizing the number of responses), known as GLM [72], in order to analyze the data obtained during the empirical validation, in which both levels for the factors (method M1 and method M2) are applied to each subject. However, prior studies [205] conclude that the existence of blocking variables (in our case, the variable P, corresponding to the archaeological problem analyzed) requires a greater treatment than that offered by the GLM model. Thus, the use of a **Mixed** statistical model with the type of covariance for "unstructured" [205] repeated measures is recommended.

Therefore, we define this model for the analysis, in which the factor (in our case the data analysis method) and the blocking variable (in our case the archaeological problem P) are defined as fixed variables as we apply two levels of both variables to all the subjects participating in the validation. The subjects are defined as a random

Table 12.4 Analysis performed about conclusion validity threats

Conclusion validity

Threat	Reason	State	Treatment of the threat
Low statistical power	The size of the sample is not big enough	Avoided	We avoid the threat by maximizing the number of subjects with repeated measures and calculating the statistical representativeness for each null hypothesis
Subjects of random heterogeneity	The subjects have not been selected at random and their profile is heterogeneous	Avoided	The subjects are selected at random from among all the volunteers. The profile is heterogeneous, albeit with a common feature: all of them are archaeologists
Fishing	The conductors of the empirical validation seek a specific result	Avoided	We avoid this threat by processing all the data gathered in order to avoid introducing bias when filtering
Reliability of measures	There is no guarantee that the results are the same when measuring the phenomenon again	Partly suffered	The metrics for accuracy, efficiency, productivity and correctness of the knowledge generated are objective The metrics for satisfaction are subjective, so they are suffered by the threat
Reliability of the implementation of the treatment	There is the risk that the application is not similar among the different people requesting the treatment or on different occasions	Partly suffered	The implementation must be as standard as possible among the different subjects and occasions. We mitigate the threat of different occasions by controlling the sessions, and of different treatments by homogenizing the possibilities of the tools used in the two treatments. In this way, we believe that the differences in application between subjects will be minimized
Random irrelevancies in experimental settings	There are elements which are external to the empirical validation which may interfere in it	Suffered	We cannot guarantee that the subject does not carry out any other task while performing the validation. We mitigate the threat by actively supervising the sessions

Table 12.5 Analysis performed about construct validity threats

Conclusion validity

Threat	Reason	State	Treatment of the threat
Interaction of testing and treatment	The subjects apply the metrics to the treatments	Avoided	The conductors of the empirical validation apply the metrics
Mono-method bias	The empirical validation with only one type of measure may present bias	Partly suffered	The variables regarding accuracy, satisfaction and quality of the knowledge generated present more than one metric, thus avoiding the threat The variables regarding efficiency and productivity suffer the threat, which we minimize by automatizing the time measurement
Hypothesis guessing	The subjects sense the purpose of the empirical validation and act in consequence	Suffered	We minimize the threat by not commenting on the objectives, research questions and defined metrics with the subjects
Evaluation apprehension	The subjects are apprehensive about being evaluated	Avoided	We avoid this threat by not commenting on the evaluative nature of the validation with the subjects and including the tasks in a research talk/visit in order to find out about their working methods
Interaction of different treatments	The result may be caused by the combination of treatments applied	Suffered	Due to the fact that the traditional method is always applied first, the validation suffers this threat. We cannot ensure that the order of application does not affect the results
Mono-operation bias	An operationalization of the treatments based on just one method may introduce bias	Suffered	Due to the fact that the data analysis methods are analyzed using specific tools (excel and the framework proposed), the validation can be affected. It may be dangerous to generalize the results to other spreadsheets. As far as the framework is concerned, it must be validated as a data analysis method. Therefore, it must be chosen

Table 12.6 Analysis performed about internal validity threats

Internal validity			
Threat	Reason	State	Treatment of the threat
History	The different treatments are applied to the same object with a significant time difference	Avoided	We avoid this threat by minimizing the time between sessions and by maintaining communication with the subjects during this time
Learning of objects	The subjects may acquire knowledge from the first treatment and apply it to the second	Avoided	We avoid this threat by using two different problems which do not allow the subjects to learn aspects of the object
Subject motivation	Less motivated subjects may present worse results than more motivated ones	Suffered	We mitigate the threat by using only volunteer subjects whose motivation in the validation is high
Maturation	The subjects react differently as time passes	Avoided	We avoid this threat by applying the traditional method first as it is the one which is normally used
Selection	The results are affected by how the subjects are selected	Avoided	The subjects are volunteers in the validation
Resentful demoralization	The subjects only apply one treatment	Not applicable	The subjects apply both treatments
Mortality	The subjects may abandon the validation before the end	Suffered	Due to the fact that the subjects are volunteers and that two sessions take place, the subjects may abandon the validation. We mitigate this threat by minimizing the time between sessions and maintaining communication with the subjects between sessions
Compensatory rivalry	The subjects who apply only the less desired treatment may influence the results	Not applicable	The subjects apply both treatments

variable, due to randomness in the process of making up the sample (as was stated earlier, an open call for participation was made to a mailing list of archaeological professionals in Spain or whose subjects were Spanish, with those offering themselves as volunteers making up our sample).

In order to be able to apply the aforementioned mixed model, we must first check if it complies with the assumption of normality of residuals. This condition can be checked by applying a K-S test to each response variable analyzed during

Table 12.7 Analysis performed about external validity threats

External validity			
Threat	Reason	State	Treatment of the threat
Interaction of selection and treatment	The subjects do not represent the general population at which the validation is aimed	Partly suffered	The subjects have different sub-profiles (or academic backgrounds) but one common feature: all of them are working in archaeology. We believe that this fact mitigates the threat in terms of generalization, although the results are only valid for similar profiles
Object dependency	The results depend on the objects used and cannot be generalized	Suffered	We minimize the threat using two objects for each treatment
Interaction of history and treatment	The treatments are applied on different days: the circumstances of the moment may affect them	Suffered	The threat is suffered due to the existence of two sessions. We minimize the threat by applying both treatments in the same room and at the same time during each session
Interaction of setting and treatment	The elements used in the validation are obsolete	Not applicable	The questionnaires used are published and in use

the application of the mixed model [72]. This test was carried out for each of our response variables, with all of them passing except the **Accuracy_TaskC** variable. For this reason, the application of the mixed model was made viable for all except the latter response variable, whose difference between the values obtained for each method is not big enough to apply a mixed model and. Therefore, we assume that this difference indicates that there are no significant levels for either of the two levels of the method.

The application of the mixed model enables us to find out whether the hypotheses made previously, and defined via the response variable, can be confirmed. In order to do this, we observe the **p-value** offered for each hypothesis to be tested, as well as the level of satisfaction offered by the model for each variable. If the level of significance α is less than 0.05, the null hypothesis must be rejected, due to the fact that there are significant differences between the two treatments. In the opposite case (α being greater than 0.05), there is no evidence to reject the null hypothesis. The application of the model and the analyses specified previously have been carried out using the SPSS V23 suite [123].

However, to what degree of magnitude is this difference significant? The **effect size** [109] enables us to know the magnitude of the differences for each factor. This

is normally only applied in cases of null hypotheses refuted beforehand via the
p-value. There are several coefficients which enable us to evaluate this effect size. In
this case, the **Cohen's d** coefficient [52, 53] has been selected in order to calculate
the effect size, due to the fact that its entries are studies of averages which are
normally carried out when two treatments are applied to one subject. Cohen's d is
defined as the difference between two averages divided by the standard deviation
which the data presents. A value of the effect size of between 0.2 and 0.49 implies a
small effect, between 0.5 and 0.79 a moderate effect while greater than 0.8 represents a large effect.

In this point, thanks to the application of the mixed model and the later measuring of the effect size, we can discover which null hypotheses are refuted and at
what magnitude. The mixed model will also offer us a calculation of the p-value
combining the response variables with the blocking variable for each hypothesis
being tested (see Problem*Method column). This enables us to know whether the
blocking variable (in our case the archaeological problem in question) is interfering
significantly in the hypothesis being tested.

Last of all, it is important to point out that the conclusions of the statistical
studies (and other studies and/or models which we could have applied) also depend
on the power of a statistical test, in other words, the probability which exists of the
refutation of a null hypothesis. This probability gives us an idea of how representative the sample taken is with regard to the total population with which we are
concerned and, therefore, what capacity of generalization we reach with the validation that we are carrying out. The application of a mixed model does not allow for
the statistical calculation of the power (as a statistical impossibility independently
of the statistical tool we may use). Due to the importance of the information
regarding the representativeness of the sample offered by this test, we have simu-
lated a statistical test of standard repeated measures (as, in our case, we have two
treatments which are applied to the same subjects) in order to calculate with the
G*Power tool [81, 82] what sample size is necessary in a model of repeated
measures in order to obtain a specific statistical power. In our case, we selected the
value generally used in order to obtain a high power (power = 0.95) and a moderate
effect size (effect size = 0.5). In order to obtain these values, a sample of at least **12
subjects** is required. Although this sample size is calculated for a model of repeated
measures without a blocking variable, we believe that our sample of 16 subjects
(and, therefore, within the magnitude of the estimate made) is adequate as, in a
model of repeated measures, it implies a moderate-high statistical power from our
empirical validation.

The following sections analyze the results obtained for the p-value and Cohen's
d for each of the defined null hypotheses.

Results and Discussion of Hypothesis H01: Accuracy

Hypothesis H_{01} stated: Accuracy in tasks relating to the generation of knowledge in archaeology using the framework proposed is similar to the accuracy obtained when performing the same tasks with the traditional method of data analysis.

This accuracy is defined by the measurement of the percentage of correct answers given by the subjects when performing the defined knowledge generation tasks. The percentages were measured for each task (from task A to task F), in order to aggregate them with two more variables:

- *Accuracy_Total_AllNothing* indicates accuracy only for the tasks in which all the sub-tasks were performed correctly.
- *Accuracy_TotalWeighted* aggregates the percentages taking into account the individual percentage of each task and giving all the same tasks the same weight.

Table 12.8 shows the *p*-values and the Cohen's d coefficients obtained for each of the tasks and for the two accuracy variables which aggregate all the tasks. Cohen's d was only calculated if the Method factor obtained significant values.

As can be seen in Table 12.8, the model offers *p*-values of less than 0.05 for accuracy in tasks **A, D, E** and **F** and in the variables which aggregate the total accuracy of the data analysis tasks taking into account only those performed correctly (***Accuracy_Total_AllNothing***) and giving the same weight to all the tasks (***Accuracy_TotalWeighted***).

For all the accuracy variables which offer significant results, the averages of the results obtained are higher for method M2 than for method M1 (see on line in [178] for the averages), which indicates better results for accuracy when using the framework proposed than when using the traditional method. The values corresponding to Cohen's d coefficient for accuracy in tasks A, D, E and F for the aggregated variables are higher than 0.8, indicating a large effect size. This can be seen more clearly in Figs. 12.19 and 12.20, which show box-and-whisker plots for

Table 12.8 *P*-values and Cohen's d for the accuracy variables

Variable	P-value			Cohen's d
	Method	Problem	Problem*Method	
Accuracy_TaskA	**0.000**	0.107	0.409	1.60
Accuracy_TaskB	1	0.249	**0.004**	
Accuracy_TaskC	–	–	–	
Accuracy_TaskD	**0.007**	0.671	0.994	1.22
Accuracy_TaskE	**0.010**	0.678	0.120	1.17
Accuracy_TaskF	**0.003**	0.312	1	1.34
Accuracy_Total_AllNothing	**0.000**	0.745	0.723	2.01
Accuracy_TotalWeighted	**0.000**	0.943	0.824	2.12

The values in bold show significant *p*-values which refute the null hypothesis H_{01}

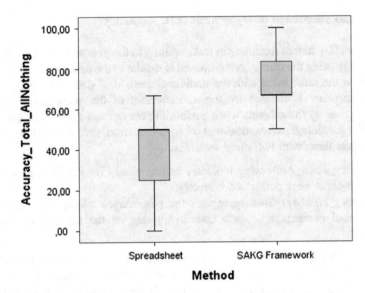

Fig. 12.19 A box-and-whisker plot for the Accuracy_Total_AllNothing variable

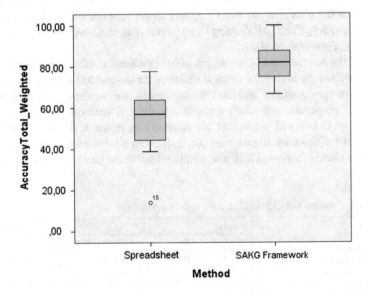

Fig. 12.20 A box-and-whisker plot for the Accuracy_Total_Weighted variable

the two aggregate accuracy variables. Box-and-whisker plots are a visual presentation which describe several important characteristics of the variable represented, such as dispersion and symmetry. They are made by representing the three quartiles and minimum and maximum values of the data on a rectangle. The longer sides show the interquartile range. This rectangle is divided by a segment which indicates the position of the median and, therefore, its relation with the first and third quartiles (it should be remembered that the second quartile coincides with the median). Any atypical values are represented by small circles outside of the central rectangle. In this case, we can observe how, in both cases, the median, the first and third quartiles for accuracy using the SAKG framework is greater than when using the traditional method with spreadsheets. This means that the subjects obtained better results in terms of aggregate accuracy when working with the framework compared to the traditional method.

As far as the remaining tasks are concerned, there are no significant results for task B relating to grouping processes. In this case, the Problem*Method interaction offers a significant result, which means that our blocking variable P (the archaeological problem being treated) is affecting the treatment (our Method). Figure 12.21 shows a profile graph of the interaction produced in the values for task B. It can be seen how, in this case, the method is only significant for one of the problems (in this case the problem of A Romea). Therefore, the subjects obtained significantly better results in terms of accuracy for task B only in the case of A Romea.

As far as the results for task C are concerned, it has not been possible to obtain results for the mixed model. This is due to the fact that, when applying the model to the values obtained for that task, a prior application criteria was not fulfilled (the residuals do not present a normal distribution). This means that the model does not

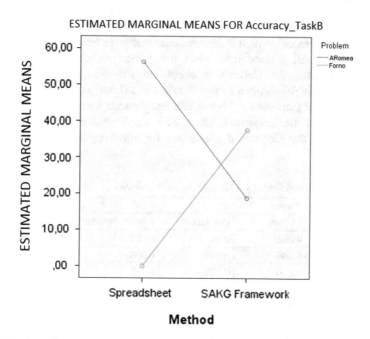

Fig. 12.21 A profile graph showing the Method*Problem interaction for the Accuracy_TaskB variable

show sufficient differences between the two treatments to offer results which guarantee the validity of the adjustment to the model. This can be seen by observing the accuracy values obtained for task C in [178], which are very similar between the two methods (almost all of the subjects carried out task C correctly, independently of the method used).

Results and Discussion of Hypothesis H02: Efficiency

Hypothesis H_{02} stated: Efficiency in tasks relating to the generation of knowledge in archaeology using the framework proposed is similar to efficiency when performing the same tasks with the traditional method of data analysis.

This efficiency has been defined by the measurement of the response time when the subjects carry out these tasks. Measurements of the time taken for each task (from task A to task F) were taken, along with measurements of the total time employed.

Table 12.9 shows the p-values and Cohen's d coefficients obtained for each of the tasks and for the total efficiency. Cohen's d has only been calculated if the Method factor obtains significant values.

As can be seen in Table 12.9, the model offers p-values of less than 0.05 for efficiency in tasks **A, D, E** and **F**, and for the **Effort_T** variable, which aggregates the total efficiency in data analysis tasks.

For all the accuracy variables which offer significant results, the average of the results obtained are higher for method M1 than for method M2 (see on line in [178] to consult the averages), which indicates better results in terms of efficiency when using the traditional method than when using the framework proposed. If we examine each task, the Cohen's d coefficient for efficiency in task E (the **Time_TaskE** variable) shows a moderate effect size (between 0.5 and 0.75), which implies that the subjects obtained better efficiency results for the traditional method in task E than for the framework, albeit with a moderate difference. The values corresponding to the Cohen's d coefficient for efficiency in the **Time_TaskA**,

Table 12.9 P-values and Cohen's d for the efficiency variables

Variable	P-value			Cohen's d
	Method	Problem	Problem*Method	
Time_TaskA	**0.002**	*0.005*	0.087	1.00
Time_TaskB	0.206	0.360	0.078	
Time_TaskC	0.870	0.088	0.212	
Time_TaskD	**0.018**	1	0.142	0.92
Time_TaskE	**0.049**	0.172	*0.012*	0.77
Time_TaskF	**0.000**	0.844	0.239	1.29
Effort_T	**0.000**	0.078	0.074	1.45

The values in bold show significant p-values which refute the null hypothesis H_{02}

Time_TaskD, Time_TaskF variables and the aggregate variable **Effort_T** are higher than 0.8, which indicates a large effect size.

Figure 12.22 shows the box-and-whisker plot for the aggregate efficiency variable. It can be observed how the median and the first and third quartiles for efficiency using the traditional method are greater than when using the SAKG framework. This means that the subjects obtained better results in terms of efficiency (they improved their response times) when working with the traditional method than when using the framework.

However, it should be noted that the graph in Fig. 12.22 shows atypical efficiency values for method M2, indicating subjects whose efficiency values using the SAKG framework in some cases (see the raw data on line in [178]) equaled or even improved on those obtained with the traditional method. In spite of the fact that, in general, the subjects presented better results in terms of efficiency with the traditional method than with the SAKG framework, it has to be taken into account that all the subjects have prior expertise in the traditional method, whereas no prior training was provided for the framework. Although this will be discussed later, we believe that the difference in the level of expertise with the method applied may be a determining factor in the results obtained in terms of efficiency and that these atypical values arise in subjects who found the framework more intuitive and, therefore, were able to overcome the expertise barrier.

To conclude the analysis, although the model offers significant values in terms of the method in efficiency of task A, it must be highlighted that it also offers significant values for the Problem, which means that there are differences in this task between subjects who worked with P1 (A Romea) and P2 (Forno dos Mouros).

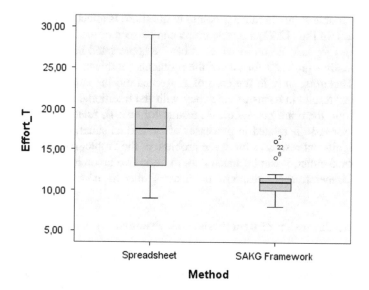

Fig. 12.22 A box-and-whisker plot for the Effort_T variable

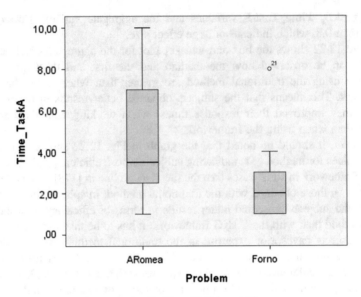

Fig. 12.23 A box-and-whisker plot for the Time_TaskA variable

Figure 12.23 shows the box-and-whisker plot in efficiency for task A, in which we can observe that the subjects evaluating problem P1 obtained better results than those evaluating problem P2.

Furthermore, it is necessary to highlight the fact that, for the **Time_TaskE** variable, related to processes of analyzing the internal structure of the data, the Problem*Method interaction offers a significant result, which means that our blocking variable P (the heritage problem in question) is affecting the treatment (of our Method). In Fig. 12.24, a profile graph can be seen of the interaction produced in the values for task E and in which it can be appreciated how, in this case, the method is only significant for one of the problems (in this case, the problem of A Romea). Therefore, only in the case of A Romea did the subjects obtain significantly better results in terms of efficiency with the traditional method for task E.

Last of all, there are no significant results for task B, related to grouping processes, or for task C, related to processes of contextual situation. Given that there were no significant results as far as the problem or the Problem*Method interaction are concerned either, it can be stated that, in these two tasks, hypothesis H_{02} is not refuted and, therefore, the response times are similar for both methods.

Results and discussion of hypothesis H03: Productivity

Hypothesis H_{03} stated: Productivity in tasks relating to the generation of knowledge in archaeology using the framework proposed is similar to productivity when performing the same tasks with the traditional method of data analysis.

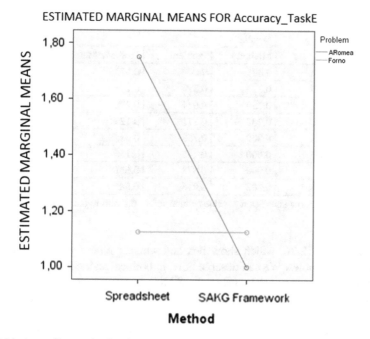

Fig. 12.24 A profile graph showing the Method*Problem interaction for the Time_TaskE variable

This productivity has been defined as the ratio of the accuracy obtained by the response time employed in performing the indicated data analysis tasks. Productivity was measured for each task (from task A to task F), in order to be able to aggregate this productivity with two more variables: *Productivity_AllN* (*AllNothing*), which indicates productivity only for tasks in which all the sub-tasks were carried out correctly and *ProductivityWeighted*, which aggregates productivity taking into account the individual percentage of each task and giving it the same weight in all the tasks.

Table 12.10 shows the *p*-values and the Cohen's d coefficients obtained for each of the tasks and for the two aggregated productivity variables.

As can be seen in Table 12.10, the model offers *p*-values of less than 0.05 for productivity in tasks **A, D, E** and **F** and in the variables which aggregate the total productivity of the data analysis tasks, taking into account only those performed correctly (*Productivity_AllN*) and giving it the same weight in all the tasks (*ProductivityWeighted*).

For all the productivity variables which offer significant results, the average of the results obtained are higher for method M2 than for method M1 (see [178] to consult the averages), which indicates better results in terms of productivity when using the framework rather than the traditional method. The values corresponding to Cohen's d for accuracy in tasks A, D, E and F and for the aggregate variables are greater than 0.8, which indicates a large effect size. This can be seen more clearly in

Table 12.10 *P*-values and Cohen's d for the productivity variables

Variable	P-value			Cohen's d
	Method	Problem	Problem*Method	
Productivity_TaskA	**0.001**	*0.042*	0.679	1.59
Productivity_TaskB	0.327	0.379	0.151	
Productivity_TaskC	0.739	*0.014*	0.278	
Productivity_TaskD	**0.002**	0.171	0.125	1.32
Productivity_TaskE	**0.009**	0.758	0.088	1.20
Productivity_TaskF	**0.000**	0.145	0.156	1.98
Productivity_AllN	**0.000**	0.878	0.631	2.40
ProductivityWeighted	**0.000**	0.916	0.385	2.34

The values in bold show significant *p*-values which refute the null hypothesis H_{03}

Figs. 12.25 and 12.26, which show box-and-whisker plots for the two aggregate productivity variables. We can observe how, in both cases, the median, the first and the third quartile for accuracy using the SAKG framework are greater than when using the traditional method with spreadsheets. This means that, with the exception of one subject, who gave an atypical value in the case of the *ProductivityAllN* variable, the subjects obtained better results in terms of productivity in an aggregate way working with the framework compared with the traditional method.

Furthermore, although the model offers significant values as far as the productivity method for task A is concerned, it must be highlighted that it also offers

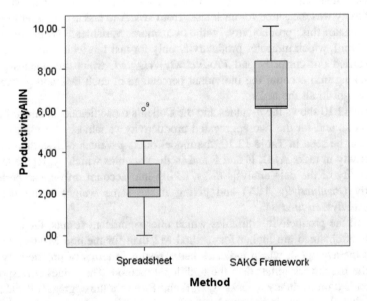

Fig. 12.25 A box-and-whisker plot for the ProductivityAllN variable

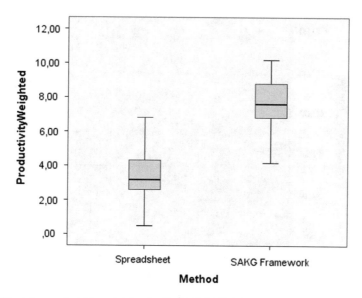

Fig. 12.26 A box-and-whisker plot for the ProductivityWeighted variable

significant values for the Problem, which means that there are differences in this task between the subjects who worked with P1 (A Romea) and P2 (Forno dos Mouros). Figure 12.27 shows a box-and-whisker plot for productivity for task A, in which it can be observed that the subjects evaluating problem P2 obtained better results than those evaluating problem P1.

Finally, there were no significant results for task B, relating to grouping processes, or for task C, relating to processes of contextual situation. In the case of task B, neither were any significant results obtained in terms of the problem or the Problem*method interaction. Thus, we can affirm that hypothesis H_{03} is not refuted and, therefore, productivity is similar with both methods. Likewise, hypothesis H_{03} is not refuted for task C, although in this case, the model offers significant values for the Problem, which means that there are differences in this task between the subjects who worked with problem P1 (A Romea) and P2 (Forno dos Mouros). Figure 12.28 shows a box-and-whisker plot for task C, in which we can observe that the subjects evaluating problem P1 obtained better results than those evaluating problem P2.

This particular issue in tasks A and C, in which the Problem is also significant without presenting interaction with the method, could indicate a degree of asymmetry between the two problems in these specific tasks.

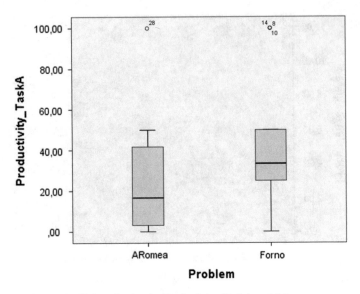

Fig. 12.27 A box-and-whisker plot for the Productivity_TaskA variable

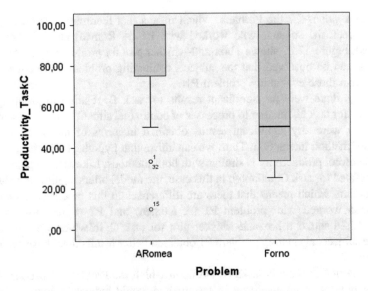

Fig. 12.28 A box-and-whisker plot for the Productivity_TaskC variable

Results and Discussion of Hypothesis H04: Satisfaction

Hypothesis H_{04} stated: The satisfaction shown by the subjects carrying out tasks relating to the generation of knowledge in archaeology using the framework

Table 12.11 *P*-values and Cohen's d for the satisfaction variables

Variable	P-value			Cohen's d
	Method	Problem	Problem*Method	
PEOU	**0.002**	0.618	0.981	1.42
PU	**0.001**	0.671	0.955	1.50
ITU	**0.000**	0.814	0.700	1.51

The values in bold show significant *p*-values which refute the null hypothesis H_{04}

proposed is similar to the satisfaction which they show when performing the same tasks with the traditional method of data analysis.

This satisfaction has been defined by measuring the scores given by the subjects according to a Likert scale dealing with three aspects: perceived usefulness (PU), perceived ease of use (PEOU) and intention to use (ITU).

Table 12.11 shows the *p*-values and the Cohen's d coefficients obtained for satisfaction in terms of PU, PEOU and ITU.

As can be seen in Table 12.11, the model offers *p*-values of less than 0.05 for values regarding perceived usefulness (PU), perceived ease of use (PEOU) and intention to use (ITU).

For all the satisfaction variables evaluated, the averages of the results obtained are higher for method M2 than for method M1 (see [178] for the averages), which indicates better results in the three aspects relating to satisfaction (perceived usefulness, perceived ease of use and intention to use) when using the framework compared to the traditional method. If Cohen's d is applied, all the values are greater than 0.8, which indicates a large effect size. This can be seen more clearly in Figs. 12.29, 12.30 and 12.31, which show box-and-whisker plots for the three variables mentioned. We can observe how, in all cases, the median, the first and the third quartile for the satisfaction variables using the SAKG framework are greater than when using spreadsheets. This means that archaeologists expressed greater satisfaction in the three criteria when evaluating the framework compared to the traditional method.

Results and Discussion of Hypothesis H05: The Quality of the Generated Knowledge

Hypothesis H_{05} stated: The degree of correctness and satisfaction shown by the subjects when evaluating the generated knowledge as a product (a written report) of the process of data analysis in archaeology using the framework proposed is similar to the degree of correctness and satisfaction which is shown when carrying out the same evaluation with the traditional method of data analysis.

This quality refers to two specific characteristics: the correctness of the generated knowledge, measured as the number of errors reported by archaeologists when analyzing the written report of another partner according to the method of data

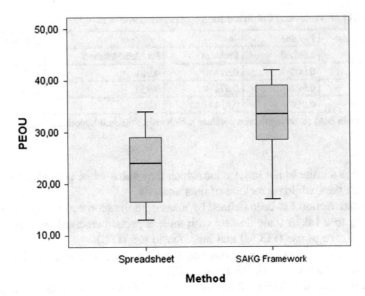

Fig. 12.29 A box-and-whisker plot for the PEOU variable

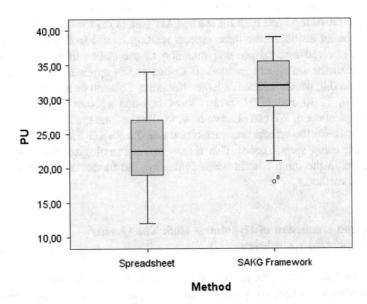

Fig. 12.30 A box-and-whisker plot for the PU variable

analysis used to produce it and the satisfaction shown by the subject with the written report, measured by the score obtained when evaluating the report via 19 sentences in a Likert scale questionnaire (see next section on this chapter for the aspects dealt with in the questionnaire).

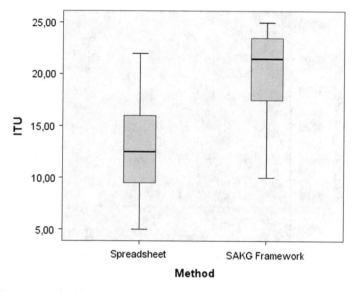

Fig. 12.31 A box-and-whisker plot for the ITU variable

Table 12.12 shows the *p*-values and Cohen's d coefficients obtained for each of the defined correctness and satisfaction variables.

As can be seen in Table 12.12, the model offers *p*-values of less than 0.05 for the variables relating to the quality of the knowledge generated. On the one hand, the **Satisfaction_KnowledgeGen** variable measures the score obtained from the Likert questionnaire. The subjects completed this questionnaire when evaluating another colleague's written reports regarding the case studies, after analyzing the data with both methods. On the other hand, the **ReportErrors_Total** variable analyses the number of errors identified by the subjects when evaluating each report, drawing up a ratio of errors according to the analysis method used to produce the report.

In the case of the **Satisfaction_KnowledgeGen** variable, the averages of the results obtained are higher for method M2 than for method M1 (see [178] to consult the averages), indicating better results in terms of satisfaction when evaluating the written reports produced using the framework proposed than those produced using the traditional method. The value corresponding to the Cohen's d coefficient for the **Satisfaction_KnowledgeGen** variable is greater than 0.8, which indicates a large

Table 12.12 *P*-values and Cohen's d for the variables relating to the knowledge generated

Variable	*P*-value			Cohen's d
	Method	Problem	Problem*Method	
Satisfaction_KnowledgeGen	**0.000**	0.718	0.476	0.99
ReportErrors_Total	**0.005**	0.680	0.639	0.85

The values in bold show significant *p*-values which refute the null hypothesis H_{05}

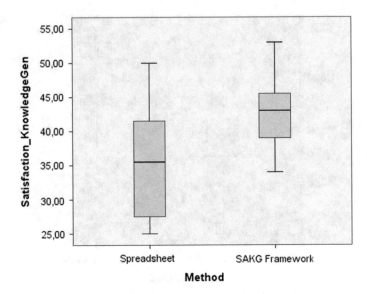

Fig. 12.32 A box-and-whisker plot for the Satisfaction_KnowledgeGen variable

effect size. This can be seen more clearly in Fig. 12.32, which shows the box-and-whisker plot for this variable. We can observe how the median, the first quartile and the third quartile when evaluating the report produced using the SAKG framework are higher than when evaluating the report produced using spreadsheets. This means that the subjects expressed more satisfaction when evaluating reports produced after the analysis using the framework than when evaluating reports produced following an analysis with the traditional method.

As far as the **ReportErrors_Total** variable is concerned, the averages of the results obtained are higher for method M1 than for method M2 (see [178] to consult the averages), indicating a higher ratio of errors when evaluating the written reports produced using the traditional method than those produced using the framework. The value of Cohen's d for the **ReportErrors_Total** variable is greater than 0.8, which indicates a large effect size. This can be seen more clearly in Fig. 12.33, which shows the box-and-whisker plot for this variable. We can observe how the median, the first and the third quartile for **ReportErrors_Total** using the traditional method are greater than when using the SAKG framework. This means that the subjects identified a significantly higher number of errors in the reports produced following an analysis using the traditional method than in the reports produced following an analysis with the framework.

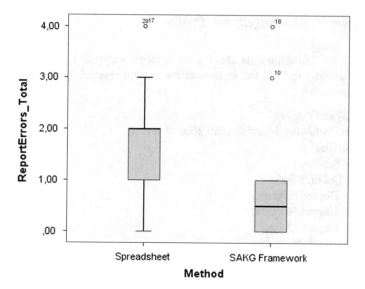

Fig. 12.33 A box-and-whisker plot for the ReportErrors_Total variable

Presentation and Package

This chapter consists of the report of the presentation of the empirical validation carried out, in terms of design and execution, the analysis and later interpretation of the results obtained and the initial implications of these results in the validation of the framework presented throughout this book. Also, it could be used as a methodological guide for any reader who wants to test their own applications created following the framework proposed in an empirical way. Next section offers our readers the original questionnaires and templates used in this empirical validation, in order to serve as a reference point for this methodological guide. The raw data obtained over the course of the empirical validation, as well as the full results of the statistical analyses carried out using the SPSS V23 suite, can be consulted on-line in [178].

Empirical Validation Materials

This section presents all the materials created and employed for the empirical validation carried on.

Questionnaire: Demographic Profile

Please, fill the following data about your demographic profile. Note that all data provided are only used in the context of the current research and processed in an anonymous manner.

Gender (Male/Female):   **Date of birth**:

Education (following Spanish educational levels):

Primary Studies

Secondary Studies

Technical Qualifications

University Degree (3 years)

University Degree (4 years)

University Degree (5 years)

M.Sc. or equivalent.

Ph.D.

Education degree title or discipline:

_____

Main discipline related to his current job:

_____

Main sector related to his current job:

(a) Public administrations (b) Private sector (c) NGOs (d) Self-employment/freelance (e)Unemployment (f)Other_____

Main function of his current job:

(a) Management (b) Teaching (c) Research activities (d) Technical activities related to my field (e) Other_____

Years of experience in the discipline related to his current job:

_____

Years of experience in curating, management, documentation, research or similar activities with archaeological data:

_____

Since your first job, what percentage of your activities have been related to interpretation of archaeological data (not only extraction and storage)?

(a) Less than 25%,

(b) Between 25 and 50% of my working experience.

(c) Between 50 and 75% of my working experience.

(d) More than 75% of my working experience.

Problem I. A Romea: Data Analysis Tasks

The A Romea case study emerges during the archaeological works performed by the company Ambiotec S.L. between December of 1999 and February of 2000. The

archaeological team documented the A Romea hillfort, an unpublished archaeological site. The following data that you can explore through Excel spreadsheets or
through the iOS prototype of the software-assisted knowledge generation framework belongs to the A Romea excavation remains and research results obtained.
Please, explore the data following the method I or II and ask the questions annotating your start and end time:

TASK A: Please, enter here the total number of fragments found in the A Romea
site in function of the material associated to them. In addition, enter also the total
number of fragments found in the A Romea site in function of the excavation
method performed.

START TIME:	END TIME:

TASK B: Please, enter here the average fragmentation percentage of the ceramic
objects found in the A Romea site and extract by a machine excavation method.

START TIME:	END TIME:

TASK C: Please, enter here the different temporal values assigned to the objects
found in the A Romea site.

START TIME:	END TIME:

TASK D: Please, enter here the stratigraphic units defined in the A Romea site
and the criterion used to group these stratigraphic units. In addition, enter also what
stratigraphic units belong to the "Barrow's chamber" group and the "Alteration
path" group.

START TIME:	END TIME:

TASK E: Please, enter here the attributes (important features documented by the
archaeological team) of the metallic object fragments found in the A Romea site.

START TIME:	END TIME:

TASK F: Please, enter here the temporal intervals associated to the A Romea's barrow at a general level having into account all data presented about the stratigraphic study and the objects and fragment found in the site. In addition, indicate the name of the tumular phase associated to Final Neolithic stage: What's happened in A Romea in that period?

START TIME:		END TIME:	

TASK G: Please, ask the following question through a textual abstract (300–500 words): What are the archaeological conclusions that you can obtain after exploring the A Romea data?

START TIME:		END TIME:	

TASK H: Please, enter here the mistakes and inconsistencies that you can identify in the abstract provided to you for evaluation, following the guidelines document attached.

START TIME:		END TIME:	

Thank you for your collaboration.

Problem II. Forno Dos Mouros: Data Analysis Tasks

The Forno dos Mouros case study emerges during the archaeological works performed in the Serra de Bocelos (Spain) between June of 1999 and July of 1999. The archaeological team documented and studied (The site was discovered previously in past campaigns) the Forno dos Mouros archaeological site. The following data that you can explore through Excel spreadsheets or through the iOS prototype of the software-assisted knowledge generation framework belongs to the Forno dos Mouros excavation remains and research results obtained. Please, explore the data following the method I or II and ask the questions annotating your start and end time:

TASK A: Please, enter here the total number of fragments found in the Forno dos Mouros site in function of the material associated to them. In addition, enter also the total number of fragments found in the site in function of the decoration characteristics of them.

START TIME: END TIME:

TASK B: Please, enter here the average fragmentation percentage of the ceramic objects found in the Forno dos Mouros site and extract by manual excavation method.

START TIME: END TIME:

TASK C: Please, enter here the different temporal values assigned to the ceramic fragments found in the Forno dos Mouros site. In addition, think about the temporal value of the ceramic object composes by these fragments: Is it possible to associate a temporal value for each ceramic object? Indicate the temporal value associated to PZ01 object and PZ06 object:

START TIME: END TIME:

TASK D: Please, enter here the stratigraphic units defined in the Forno dos Mouros site and the criterion used to group these stratigraphic units. In addition, enter also what stratigraphic units belong to the "First megalithic zone" group and the "Chamber access: pit" group.

START TIME: END TIME:

TASK E: Please, enter here the attributes (important features documented by the archaeological team) of the stratigraphic unit defined in the Forno dos Mouros site.

START TIME: END TIME:

TASK F: Please, enter here the temporal intervals associated to the Forno dos Mouros site at a general level, having into account all data presented about the stratigraphic study and the objects and fragment found in the site. In addition, indicate the name of the tumular phase associated to Initial Neolithic stage: What's happened in Forno dos Mouros in that period?

| START TIME: | END TIME: |

TASK G: Please, ask the following question through a textual abstract (300–500 words): What are the archaeological conclusions that you can obtain after exploring the Forno dos Mouros data?

| START TIME: | END TIME: |

TASK H: Please, enter here the mistakes and inconsistencies that you can identify in the abstract provided to you for evaluation, following the guidelines document attached.

| START TIME: | END TIME: |

Thank you for your collaboration.

Questionnaire Data-Analysis Method I: Spreadsheets in Microsoft Excel®

The questionnaire gives you the possibility to express your opinion about the use of Excel spreadsheets to perform data analysis tasks during the knowledge generation process. Note that a data analysis task is one that is carried out to examine raw data in order to extract knowledge and achieve some conclusions. Please read each sentence carefully and give a score according to the following values: 1 = Totally Disagree, 2 = Moderately Disagree, 3 = Neutral, 4 = Moderately Agree, 5 = Totally Agree.

Sentences	1	2	3	4	5
1. I think the procedure for data analysis using Excel spreadsheets k is simple and easy to follow	O	O	O	O	O
2. I think Excel spreadsheets reduce the effort I need to do to perform the required data analysis tasks	O	O	O	O	O
3. Excel spreadsheets allow the grouping of the data following different criteria in an agile and intuitive manner	O	O	O	O	O
4. Using Excel spreadsheets I can efficiently perform data grouping following my own criteria	O	O	O	O	O

(continued)

(continued)

Sentences	1	2	3	4	5
5. Excel spreadsheets allow me to combine data values in order to extract conclusions in an agile and intuitive manner	O	O	O	O	O
6. Using Excel spreadsheets I can efficiently perform exploratory data analysis tasks that relate different dimensions (geographical, temporal and provenance), and test different hypotheses	O	O	O	O	O
7. Excel spreadsheets allow me to explore contextual dimensions (geographical, temporal or provenance) of my data in an agile and intuitive manner	O	O	O	O	O
8. Using Excel spreadsheets I can know more about the contextual dimensions (geographical, temporal or provenance) of my data, which allows me to answer questions about their structure	O	O	O	O	O
9. Excel spreadsheets allow me to know more about the internal structure of my data (i.e. the underlying schema) in an agile and intuitive manner	O	O	O	O	O
10. Using Excel spreadsheets I can know more about the internal structure of my data (i.e. the underlying schema), which allows me to answer questions about it	O	O	O	O	O
11. In general, I find Excel spreadsheets easy to use	O	O	O	O	O
12. I think I could easily explain how to perform data analysis tasks using Excel spreadsheets to others	O	O	O	O	O
13. In general, I find Excel spreadsheets helpful	O	O	O	O	O
14. In general, Excel spreadsheets could be a practical method to perform data analysis tasks and obtain conclusions	O	O	O	O	O
15. In my opinion, it is easy to use Excel spreadsheets to perform the data analysis tasks required in the specific case study that was analyzed	O	O	O	O	O
16. I would use Excel spreadsheets if it was available to group my data according to my own criteria	O	O	O	O	O
17. I would use Excel spreadsheets if it was available to combine data values and extract some conclusions	O	O	O	O	O
18. I would use Excel spreadsheets if it was available to explore contextual dimensions (geographical, temporal or provenance) of my data	O	O	O	O	O
19. I would use Excel spreadsheets if it was available to know more about the internal structure (i.e. the underlying schema) of my data	O	O	O	O	O
20. I feel I have the necessary skills to perform data analysis tasks with Excel spreadsheets	O	O	O	O	O
21. In general, I think Excel spreadsheets offers significant improvements over other methods to perform data analysis tasks in research and knowledge-generation	O	O	O	O	O
22. I will try to use Excel spreadsheets if I had to perform data analysis tasks in research and knowledge-generation in the future	O	O	O	O	O

Questionnaire Data-Analysis Method II: Software-Assisted Knowledge Generation Framework

The questionnaire gives you the possibility to express your opinion about the use of our software-assisted knowledge generation framework to perform data analysis tasks during the knowledge generation process. Note that a data analysis task is one that is carried out to examine raw data in order to extract knowledge and achieve some conclusions. Please read each sentence carefully and give a score according to the following values: 1 = Totally Disagree, 2 = Moderately Disagree, 3 = Neutral, 4 = Moderately Agree, 5 = Totally Agree.

Sentences	1	2	3	4	5
1. I think the procedure for data analysis using the software-assisted knowledge generation framework is simple and easy to follow	O	O	O	O	O
2. I think the software-assisted knowledge generation framework reduces the effort I need to do to perform the required data analysis tasks	O	O	O	O	O
3. The software-assisted knowledge generation framework allows the grouping of the data following different criteria in an agile and intuitive manner	O	O	O	O	O
4. Using the software-assisted knowledge generation framework I can efficiently perform data grouping following my own criteria	O	O	O	O	O
5. The software-assisted knowledge generation framework allows me to combine data values in order to extract conclusions in an agile and intuitive manner	O	O	O	O	O
6. Using the software-assisted knowledge generation framework I can efficiently perform exploratory data analysis tasks that relate different dimensions (geographical, temporal and provenance), and test different hypotheses	O	O	O	O	O
7. The software-assisted knowledge generation framework allows me to explore contextual dimensions (geographical, temporal or provenance) of my data in an agile and intuitive manner	O	O	O	O	O
8. Using the software-assisted knowledge generation framework I can know more about the contextual dimensions (geographical, temporal or provenance) of my data, which allows me to answer questions about their structure	O	O	O	O	O
9. The software-assisted knowledge generation framework allows me to know more about the internal structure of my data (i.e. the underlying schema) in an agile and intuitive manner	O	O	O	O	O
10. Using the software-assisted knowledge generation framework I can know more about the internal structure of my data (i.e. the underlying schema), which allows me to answer questions about it	O	O	O	O	O

(continued)

(continued)

Sentences	1	2	3	4	5
11. In general, I find the software-assisted knowledge generation framework easy to use	O	O	O	O	O
12. I think I could easily explain how to perform data analysis tasks using the software-assisted knowledge generation framework to others	O	O	O	O	O
13. In general, I find the software-assisted knowledge generation framework helpful	O	O	O	O	O
14. In general, the software-assisted knowledge generation framework could be a practical method to perform data analysis tasks and obtain conclusions	O	O	O	O	O
15. In my opinion, it is easy to use the software-assisted knowledge generation framework to perform the data analysis tasks required in the specific case study that was analyzed	O	O	O	O	O
16. I would use the software-assisted knowledge generation framework if it was available to group my data according to my own criteria	O	O	O	O	O
17. I would use the software-assisted knowledge generation framework if it was available to combine data values and extract some conclusions	O	O	O	O	O
18. I would use the software-assisted knowledge generation framework if it was available to explore contextual dimensions (geographical, temporal or provenance) of my data	O	O	O	O	O
19. I would use the software-assisted knowledge generation framework if it was available to know more about the internal structure (i.e. the underlying schema) of my data	O	O	O	O	O
20. I feel I have the necessary skills to perform data analysis tasks with the software-assisted knowledge generation framework	O	O	O	O	O
21. In general, I think the software-assisted knowledge generation framework offers significant improvements over other methods to perform data analysis tasks in research and knowledge-generation	O	O	O	O	O
22. I will try to use the software-assisted knowledge generation framework if I had to perform data analysis tasks in research and knowledge-generation in the future	O	O	O	O	O

Guidelines for Evaluating Textual Abstracts: Possible Mistakes

This document presents a set of guidelines for evaluating the textual abstract created in the context of the current empirical validation. The guidelines attend to the main mistakes and inconsistencies that the abstract could contain. Please, read the abstract carefully and identify the following mistakes and inconsistencies, by answering these questions:

- Explicit references to numeric data values presented in the case study analyzed. For instance, "30 ceramic fragments are recovered which only 3 of them correspond to vessel's rims".

  - How many similar references appears in the abstract?

  _____

  - How many similar references presented in the abstract are mistakes?

  _____

- Sentences containing references to processes performed by the author to obtain conclusions from the data, for instance, use of verbs such as compare, classify, identify, categorize, etc. or causalities, exemplifications and generalizations.

  - How many similar references appears in the abstract?

  _____

  - How many of them you think are well-founded by the author based on the data of the case study that you analyzed?

  _____

  - How many of them you think you would obtain the same conclusion than the author?

  _____

- References with redundancies referring to a specific aspect in the abstract, such as repetition of argumentation, same references to numerical data of the case or repetition of examples.
- How many similar references appears in the abstract?

  _____

- Sentences containing references to conclusions of the author about the case study:
- How many similar references appears in the abstract?

  _____

- How many of them you think are well-founded by the author based on the data of the case study that you analyzed?

  _____

- How many of them you think you would obtain the same conclusion than the author?

  _____

Thank you for your collaboration.

Evaluation Questionnaire: Satisfaction Degree About Knowledge Generation Quality

The questionnaire gives you the possibility to express your opinion about the quality of the textual abstracts generated throughout the empirical validation process. These memories are generated after performing data analysis tasks with Excel spreadsheets and the software-assisted knowledge generation framework. Please read the textual abstract carefully and give a score according to the following values: 1 = Totally   Disagree,   2 = Moderately   Disagree,   3 = Neutral, 4 = Moderately Agree, 5 = Totally Agree.

Sentences	1	2	3	4	5
1. In general, the abstract presents a structured and clear discourse organization	O	O	O	O	O
2. In general, I think the abstract summarizes the case study data analyzed	O	O	O	O	O
3. The abstract presents redundancies in data references and motivation reasoning	O	O	O	O	O
4. I can clearly identify the research conclusion specified in the abstract	O	O	O	O	O
5. I think the conclusion expressed in the abstract—explicitly or implicitly— is well-founded based on the data	O	O	O	O	O
6. The abstract contains references to the raw data used as a basis for reasoning	O	O	O	O	O
7. The abstract contains references to the relation between the raw data analyzed and the research conclusion achieved	O	O	O	O	O
8. In my opinion, the abstract contains inconsistencies or contradictions	O	O	O	O	O
9. After reading the abstract, I know what kinds of data have been analyzed and its internal structure	O	O	O	O	O
10. I think there is a lack of references to important data of the case study in the abstract, which should be referenced for the complete understanding of it	O	O	O	O	O
11. The abstract contains references to what processes have been performed to analyze the data: grouping, comparing, etc	O	O	O	O	O
12. After reading the abstract, I can give a better answer to the question: What are the conclusions achieved for the authors from the data?	O	O	O	O	O
13. After reading the abstract, I can give a better answer to the question: How the conclusions have been generated?	O	O	O	O	O
14. After reading the abstract, I know how the author is reasoning about the data of the case study	O	O	O	O	O
15. I think the abstract is a good synopsis about the research performed based on the case study analyzed	O	O	O	O	O

(continued)

(continued)

Sentences	1	2	3	4	5
16. I think the abstract allows me having an overview of the case study analyzed	O	O	O	O	O
17. I think the abstract allows me creating new research questions and elaborating hypotheses about the data if the case study and the conclusions achieved by the author	O	O	O	O	O
18. I think if I read the abstract without knowing the case study, I can understand the main ideas and research conclusion	O	O	O	O	O
19. In general, I understand the content of the abstract	O	O	O	O	O

Thank you for your collaboration.

Conclusions

In summary, the null hypotheses formulated have been refuted for almost all of the variables analyzed (with the exception of some, which measure variables for specific tasks). It should be highlighted that they have been refuted for all the response variables which measure tasks in an aggregate manner, even for both methods of task aggregation, taking into account the tasks performed with full accuracy and productivity and those which take into account intermediate states of accuracy and productivity.

However, this does not mean that our framework presents better results in all of the response variables which have been measured. The SAKG framework presents far better results in terms of accuracy, productivity, satisfaction of the subjects in terms of the three variables involved (perceived usefulness (PU), perceived ease of use (PEUO) and intention to use (ITU)) and in terms of the quality of the generated knowledge in the variables analyzed concerning correctness (*ReportErrors_Total*) and satisfaction (*Satisfaction_KnowledgeGen*).

Due to the fact that some variables were also evaluated for each task performed, we can carry out a more in-depth analysis of which tasks have provided better results in each response variable tested. This will allow us to gain an initial perspective of the assistance offered to the generation of knowledge and which cognitive processes have been assisted satisfactorily and which have not. Below, we shall offer detailed conclusions for each response variable dealt with.

As far as accuracy is concerned, the subjects (our archaeologists) clearly improved their rate of correct answers when using the SAKG framework for tasks relating to processes of combining values (tasks A and D), in addition to those concerning processes analyzing the internal structure of the data (task E). Furthermore, they also improved their responses in one of the tasks relating to processes of contextual situation to discover the temporal characteristics of the data (task F), although they did not do so in the other task for the same cognitive process (task C). However, accuracy was not improved for the task regarding grouping processes (task B).

In terms of productivity, the pattern observed for accuracy is repeated.

As far as efficiency is concerned though, the traditional method of using spreadsheets offered better results than the SAKG framework for tasks relating to processes of combining values (tasks A and D) and in tasks concerning processes analyzing the internal structure of the data (task E). Furthermore, the subjects also improved their response times in one task relating to processes of contextual situation to discover the temporal characteristics of the data (task F). Due to these results for efficiency, an analysis of behavior in efficiency for each task performed has been carried out. Figures 12.34 and 12.35 show box-and-whisker plots corresponding to efficiency in each task.

As can be appreciated in the graphs, there are tasks in which efficiency using the traditional method is clearly greater than with the proposed framework, for example in the cases of tasks E and F:

- Task E is related to processes of analyzing the internal structure of the data. The subjects were quicker to offer a response regarding the internal structure of the data when they were using spreadsheets than when accessing the data via the Structure IU interaction unit. On an exploratory level, we believe that the subjects reach the structure of the information more quickly when using

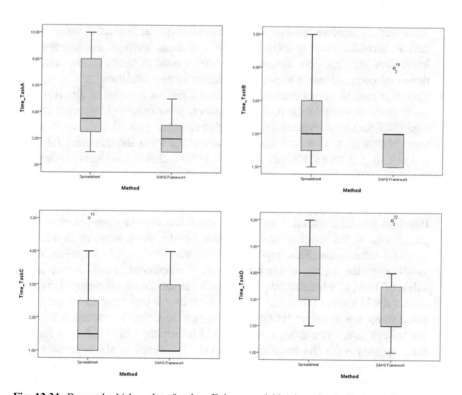

Fig. 12.34 Box-and-whisker plots for the efficiency variables in tasks A, B, C and D

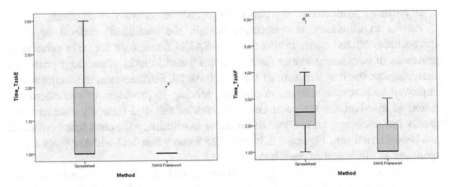

Fig. 12.35 Box-and-whisker plots for the efficiency variables in tasks E and F

spreadsheets as this structure is presented in the same interface as the data itself. In the framework proposed, they need to access the specific interaction unit via the menu in order to view the structure, which implies one more step in terms of navigation, thus slowing down access to the information. Furthermore, the difference in the level of expertise of the subjects may also have an influence in this efficiency. However, the results in accuracy and productivity for this same task were better when using the Structure IU interaction unit within the framework. Therefore, we believe that, even though the internal structure of the data is accessed more quickly with the traditional method, the Structure IU interaction unit presents significant improvements in terms of the subjects' degree of comprehension about this structure in the validation.

- Task F is related to processes of contextual situation in order to discover temporal characteristics of the data. The subjects were quicker to give an answer regarding temporal aspects of the archaeological case studies when using spreadsheets than when accessing the information via the Timeline IU interaction unit. On an exploratory level, we believe that the subjects reached the temporal information more quickly when using spreadsheets as this information is presented in a table format. The subjects search for temporal information (similar dates or data) in the spreadsheet and quickly formulate their response based on this information. However, in the framework proposed, they need to gain access to the specific interaction unit via the menu in order to view the temporal information. This implies one more step in terms of navigation, which slows down the access to this information. Furthermore, the difference in the subjects' level of expertise may also have an influence on efficiency. However, the results in terms of accuracy and productivity for this same task were better when using the Timeline IU interaction unit within the framework. Therefore, we believe that, even though the temporal information of the data is reached more quickly with the traditional method, the Timeline IU interaction unit presents significant improvements in terms of the subjects' degree of comprehension about this structure in the validation by, for example, offering a better view of the temporal phases involved in each case study.

- The results in terms of efficiency for other tasks, such as those relating to processes of combing values, present closer values between the two methods (tasks A and D). In task D, even some atypical values can be observed for subjects who presented better times with the framework. On an exploratory level, we believe, therefore, that although the subjects may present better response times when using spreadsheets compared to accessing the information via the Value-Combination IU interaction unit, this interaction unit has enabled the barrier of the subjects' prior level of expertise to be overcome and offers an acceptable level of behavior in terms of efficiency. This, together with the good results offered in this type of task in terms of accuracy and productivity, offers promising results for providing assistance to cognitive processes relating to the combination of values. This proximity between values in both methods is an aspect which can also be observed in task C, relating to processes of contextual situation in order to discover the temporal characteristics of the data.
- Task B, relating to grouping processes, presents some atypical values in favor of the framework proposed. However, neither accuracy nor productivity in this task presents significant levels implying improvements when using the framework. On an exploratory level, we believe that this task, relating to grouping processes, is the one which has shown the worst behavior in general terms in the use of the framework.

To sum up, and to repeat what has previously been pointed out, we believe that the subjects' level of familiarity with both methods and the absence of any previous training with the method based on the framework proposed (thus presenting differences between both methods for learnability [133]) may have a significant influence on this result in favor of the traditional method using spreadsheets in terms of efficiency and, therefore, it would be necessary to carry out a validation with prior training in the use of the framework proposed in order to eliminate this factor. However, with the results provided by this validation, it is necessary to take into account the fact that perhaps the framework does not improve response times compared to the traditional method, independently of the degree of familiarity of the subject with the methods being evaluated and his/her degree of expertise in using them. The remaining response variables analyzed present good results when evaluating aspects of accuracy, productivity, user satisfaction and correctness and satisfaction with the knowledge generated, when assisted by the framework proposed compared with the traditional method employed by archaeologists, based on table formats such as spreadsheets.

Both of these issues (the influence of prior familiarity with methods M1 or M2 and the difference of results between efficiency and the rest of the analyzed variables) and more detailed aspects surrounding the generation of knowledge provided and implications in specific cognitive processes will be discussed in depth in Chap. 13.

Part V
Final Considerations

Chapter 13
Discussion

Introduction

Over the course of the previous chapters, the problem of providing software assistance to the generation of knowledge based on large archaeological data sets has been explored, explained in detail and conceptualized. In addition, a full proposal offering a solution in the form of a conceptual framework has been developed. This framework consists of three parts, each with their corresponding software models (the subject model, the cognitive processes model and the presentation and interaction model), along with a metamodel in order to formally express the connections and possible interoperability between them. Finally, the readers have been guided through the application of the framework, validating how the models fits in an analytical way with the goals proposed and how built software systems using the framework presented. Also, a second empirical validation were designed and made in order to validate the proposed solution using an original prototype of a system to analyze two data archaeological case studies.

This chapter analyses the solution presented here from the point of view of the whole research process, providing answers to the research questions posed at the beginning of the book. In addition, this chapter will specifically deal, in a critical way, with areas identified, over the course of this work, as being in need of improvement and how they can be tackled.

Answering the Initial Questions

As was described in Part I, this book takes as its starting point that **it is possible to significantly improve knowledge-generation processes in archaeology by way of providing software assistance to the user with information visualization techniques**.

© Springer International Publishing AG 2018
P. Martin-Rodilla, *Digging into Software Knowledge Generation in Cultural Heritage*, Modeling and Optimization in Science and Technologies 11,
https://doi.org/10.1007/978-3-319-69188-6_13

Along the chapters we attempt to respond to this question led to the emergence of some secondary questions, making the research in progress more specific and enabling the creation of a set of techniques, tools, and methodologies explained here. Now, it's time to recapitulate the answers offered by this book to our initial questions.

First Secondary Question: Archaeological Knowledge Generation Problems

Firstly, we had the need of knowing **what problems exist in knowledge-generation processes in archaeology as they are normally carried out**.

Following a thorough study of the process of knowledge generation in archaeology through written sources, the tools employed for this aim and empirical research (described in Part II), we were able to identify seven problems, which have all been dealt with in depth in [174]. We shall outline each of them here and describe how they have been dealt with in framework proposed and applied here:

1. The intentional use of uncertainty in the intermediate reasoning to generate knowledge. This uncertainty is not supported by existing software tools.
2. The absence of explanation and, consequently, of monitoring of the cognitive process being carried out in each case, detecting, for example, cognitive processes with a strong temporal and/or spatial component. It is necessary to characterize them in order to provide software assisted knowledge generation in the field.
3. The absence of information storage concerning the tasks carried out with the information, such as which questions are asked of the data during the knowledge generation process. This impedes feedback in existing software systems, which would allow them to adapt to this process.
4. The previous point also implies an absence of support for the collaborative generation of knowledge, as it cannot be known which cognitive tasks have been carried out and which have not, thereby making group work more complicated.
5. A lack of priority management of the information on which the knowledge generated is based.
6. A lack of vision regarding the structure of the information on the part of the end users, thus making the creation and testing of hypotheses based on the data more difficult.
7. Homogeneous procedures applied to reasoning, derived from direct observation, and reasoning derived from more complex mechanisms (relation between data, abstraction, interpretation, etc.). This situation could include confusion about the level of the DIKW hierarchy [7] in which reasoning is situated and the level of subjectivity and uncertainty that is managed.

The framework presented here deals with all seven of these problems, albeit with different degrees of scope in their evolution:

- Problem 1 is dealt with in the framework via the vague support provided by ConML [129] as a language and CHARM [130] as an abstract model of reference. This forms the basis of our subject model. Its use in each specific case remains, therefore, the choice of software analyst or domain specialist, who will be able to include more or less elaborate mechanisms in order to deal with vagueness in the subject model of each type of software assistance.
- Problem 4 has only been dealt with in the framework thanks to the cognitive processes metamodel. The empirical validation in groups has enabled us to find out how the prototype we built behaves, based on the framework in multi-user contexts. It will desirable in the future also to test and adapt if necessary the framework for collaborative software tools (that is to say, ones in which the users are not only able to use the tool, but are also able to react with the results of other users at the same time).
- Problems 2, 3 and 7 are dealt with in the framework thanks to the cognitive processes metamodel, which allows one or several cognitive processes being assisted to be instanced, saving the one being assisted in each case, along with the actions being carried out by the user for a specific data subset.
- Problems 5, 6 and, to some extent, problem 2 are dealt with thanks to specific interaction units and individual patterns for the treatment of temporal and/or geographic aspects and for handling levels of importance and reasoning based on the structure of the information. All of these aspects are present in the hierarchy of presentation and interaction patterns defined in the framework.

Second Secondary Question: Archaeological Cognitive Processes

Secondly, we determined that the majority of the problems detected, in order to be assisted via software, required a characterization of the most common cognitive processes in archaeology. Thus, we wondered what **are the most common cognitive processes carried out by archaeologists in the generation of knowledge**.

In an attempt to respond to this question, characterizations of existing cognitive processes in similar fields were studied and discourse analysis techniques were applied to textual sources produced in archaeology, allowing the most common cognitive processes to be characterized into four groups of primitives: Building (inferences based on the structure, rather than on particular values, of the data), Clustering (inferences based on the grouping of the data), Situating (inferences based on situating the data in a particular context or setting) and Combining (inferences regarding the basis of the combination and/or the comparison of the values of the different elements of the data).

The details of the characterization and methodology proposed in order to integrate discourse analysis techniques into software engineering is Part II and have been published widely [171, 176]. Moreover, we described the cognitive processes metamodel present in our framework, which formalizes this characterization, in Chap. 8.

Third Secondary Question: Information Visualization Techniques for Large Archaeological Data Sets

Finally, we needed to know **which are some appropriate information visualization techniques to assist each one of the cognitive processes identified within archaeology**, in order to offer to archaeologists.

Before a wide revision of the area and performing empirical studies with archaeologist (all described in Part II), led us to define a hierarchy of patterns which encapsulated the solutions for presentation and interaction observed in the interaction units for each cognitive process. The answer is, therefore, the metamodel of the presentation and interaction mechanism described as part of the framework.

Final Discussion

Once the three secondary questions had been answered, we can reaffirm that it **is indeed possible to improve the processes of knowledge generation in archaeology by means of providing software assistance to the user via information visualization techniques**. The most importance evidences showed in this book are:

- Evidence has been observed, reflected throughout the entire work done and referred in this book, as well as through the appearance of projects, proposals, infrastructures, disciplines, etc., which establish the need for software assistance in archaeology. An excellent example of this is the appearance of *Digital Humanities* as a discipline, which leads us to believe there is a need for software assistance per se. In relation to this, the presence of certain assistance mechanisms which have already been implemented and/or conceptualized, at differing stages of development, has been detected, especially in the application of geographic information systems in the field, which would indicate that software assistance is, indeed, possible. This evidence is described in the bibliographic review carried out here, though especially in Chap. 3.
- Coincidences have been found in the generation of knowledge in archaeology in different users, which allows for the formalization of software assistance. These coincidences have been observed from the very beginning of the research process, both empirically (in the prior empirical studies carried out) and on an abstract level, as can be seen in the appearance, confirmation and later

formalization of the textual patterns which arose from the analysis of archaeological discourses.

- How well the conceptual framework presented in order to represent assistance to knowledge generation worked in the case study presented (A Romea) and in a second case study used during the empirical validation (Forno dos Mouros). This represents evidence in itself that software assistance is possible in this context.
- Further proof found is the statistically significant improvement in terms of accuracy and productivity in tasks relating to the generation of knowledge assisted by visualization techniques compared to those performed without assistance (via spreadsheets). These statistically significant improvements occur in three of the four groups of cognitive processes (Building, Situating and Combining) (see the definition of the defined data analysis tasks carried out for the empirical validation in Chap. 12).
- Likewise, a statistically significant improvement was found in terms of end-user satisfaction when carrying out tasks relating to knowledge generation which are assisted by visualization techniques compared to those performed without assistance (via spreadsheets).
- Finally, another statistically significant improvement was found in terms of the quality of the knowledge generated. The written reports produced following an analysis of data using the framework which the archaeologists then analyzed received better scores and contained fewer errors than those reports written following an analysis of data via spreadsheets.

All of this evidence and the results obtained from the validations of the proposed framework allow us to state that the proposed solution offers a type of software assistance which significantly improves knowledge generation processes based on the combination of values, the analysis of the structure of information and its contextual situation, all of which have been identified as relevant when we analyze large archaeological data sets. It does so in terms of accuracy, productivity, satisfaction and quality of the knowledge generated.

However, other aspects of interest have been detected in which the software assistance proposed as a solution has not obtained such satisfactory results, and has even presented certain restrictions. All of these aspects will be dealt with in the following section.

A Critical Analysis

First of all, there are a series of restrictions which must be taken into account when it comes to evaluating the degree to which the initial hypothesis defined at the beginning of this book is effectively supported by the solution presented. This limitation study, far from discouraging the reader in the application of the framework (that it allows to define the desirable software assistance for each archaeological data case and, as we have seen, improve the archaeologist's performance in terms of accuracy, productivity and satisfaction with the analysis), it also allows to identify those areas of research in this specific application of software-assisted knowledge generation. Thus, they represent immediate improvement opportunities. These restrictions arise from the results offered in the validations of the solution which have been carried out:

- The validations done with archaeologists demonstrates the benefits of the approach presented, although they do not present statistically significant results in terms of efficiency for the solution proposed when compared with the unassisted data analysis. As was described before, we believe that one of the factors which may have an influence in this regard is the level of prior expertise possessed by the subjects of the validation in the two methods of analysis being compared. Due to the absence of prior training with the prototype of the solution presented, the subjects possessed greater expertise in spreadsheets than with the prototype. This implies a difference in the level of familiarity between the two methods. This may have repercussions, particularly in terms of the degree of efficiency measured. However, it is possible that other factors also have an influence, such as the degree of learnability of the prototype itself not being sufficient. Note that learnability is defined according to ISO/IEC 9126, as "the capability of a software product to enable the user to learn how to use it" [133, 272]. We assume, therefore, that it is necessary to carry out larger empirical studies with differing degrees of expertise on the part of the subjects and incorporating mechanisms for prior training in order to be able to identify the factors which affect this result.
- The sample selected for the validation is statistically representative when it comes to comparing the two methods. That is to say, it is possible to generalize about the significant improvements of our proposed solution compared to the traditional method (spreadsheets). However, for an individual analysis of the proposed framework, a bigger sample, with greater heterogeneity, is necessary in order for us to not only confirm which aspects are improved compared with other methods but also to draw conclusions with a higher degree of generalization, such as whether the framework improves the quality of the written reports produced or the degree of accuracy in the tasks, without taking into account the traditional methods used for prior data analysis.

Another relevant aspect is the sphere of application of the proposed framework. First of all, the framework allows the readers to define the information visualization and interaction patterns most adequate given cognitive process to assist and the archaeological data set analyzed. However, for an entire treatment of the knowledge generation assistance that we want to provide, we believe it is necessary also deal with the knowledge extraction mechanisms, which were proposed in the initial questions and approach explained in the firsts chapters. Although this treatment is put forward as an area of future research in this book, we would like to point out that, from our point of view, software assistance integrating these mechanisms would be more complete but it would have to be designed and validated to evaluate its performance as part of the proposed framework.

Secondly, it should be emphasized that the solution proposed here has been empirically tested by archaeologists and professionals working on archaeological aspects belonging to different institutions and of different profiles. We believe that the proposed solution, apart from the methodological implications of the chosen software models, offers software assistance to these specialists, independently of their work methodology, their perspective of Cultural Heritage or archaeological background and other variables which are commonly used in archaeology, such as historical periods or chrono-cultural adscriptions, which are the specialty of the expert, etc. The capacity of the mechanisms used for the subject model (ConML and CHARM) in order to handle the models with subjectivity, temporality and vagueness, together with a high level of abstraction in the definition of the cognitive processes being assisted (Building, Combining, Clustering, Situating), allows us to make the applicability of the framework independent of these variables.

Finally, it should be pointed out that there is not only one form of software assistance and that this work takes the view that there is not only one way to reach this assistance. Therefore, it is possible that other methods or techniques, which have not been considered during this research process, may also obtain good results in terms of providing software assistance to the generation of knowledge in archaeology. Over the course of this research, an in-depth study of other prior projects was carried out, although the field in question is so broad that it is impossible to state whether this review covered one hundred per cent of the spectrum of existing studies. However, we believe that the studies examined are sufficient in terms of number and relevance to cover a broad spectrum of the state of the art in the fields in question. The flexibility of the conceptual mechanisms integrated into the framework with metamodels dealing with concepts at a high level of abstraction and hierarchies of patterns allow us to consider the incorporation, or the study along similar lines, of other models and techniques, which have not previously been considered in this research process.

The critical analysis offered here, far from minimizing the scientific contributions made, demonstrates that this research area is a work in progress, which opens up future lines of action.

Chapter 14
Conclusions

Summary

This book guides the reader for the amazing sub-area of the software-assisted knowledge generation: How it is possible through software to assist humans in creating new knowledge about our past? Specifically, the reader has found a complete conceptual framework based on software models for the development of software to assist archaeologists and related professionals in the process of knowledge generation. This chapter aims to condense the contributions presented here to the scientific community into the different disciplines and areas involved and how they can be used for the different kind of readers and practitioners. In addition, the chapter points out the emerging lines of research and future possibilities opened up by this work, offering some key about future scenarios.

We have carried out an in-depth study following a rigorous methodology—exposed in Chap. 2—of the particularities of the archaeological research domain, specially focusing on knowledge generation from large data sets. Most of the existing techniques for conceptual modeling in archaeological data sets are reviewed and discussed in Chap. 3. In Chap. 4, we used most innovative conceptual modeling strategies to integrate support systems into the framework presented for dealing with aspects such as temporality, subjectivity and vagueness in conceptual terms. At the present time, we do not know of any previous study with similar characteristics but, most important, we develop here a guide step by step for any researcher or practitioners who need represent these aspects in their data structures.

In addition, we emphasize the relevant role of the cognitive processes performed by humans when they analyzed large data sets, especially in humanistic disciplines such as archaeology. We defend the explicit modeling in software systems for this kind of information, becoming the cognitive processes modeling a key part of the software-assisted knowledge generation strategies referred here for archaeology. General principles and works from different disciplines in cognitive processes

© Springer International Publishing AG 2018
P. Martin-Rodilla, *Digging into Software Knowledge Generation in Cultural Heritage*, Modeling and Optimization in Science and Technologies 11, https://doi.org/10.1007/978-3-319-69188-6_14

modeling and interpretation are also reviewed in Chap. 3. However, cognitive processes identification requires an ad hoc study of the discipline involve and their main sources. Thus, textual sources analysis becomes an essential part of the approach adopted in order to identify the cognitive part of this research. Chapter 4 offers a complete methodology, including a modeling language based on ISO/IEC 24744, for the application of discourse analysis techniques in software engineering. We have shown how this application proves to be especially useful for those parts of the engineering process requiring the analysis of freestyle textual sources, such as requirement specifications, documents for translations of systems, etc. Furthermore, some contributions have been made in terms of the automatization and use of discourse analysis techniques for the automatic extraction of methodological information from textual sources.

Continuing with cognitive processes modeling, we did not find most reference works about their validation with real archaeologists, what methodology we must adopt and what results we should expect. Chapter 4 also offers a method to identify traces of cognitive processes which are present in textual sources and validate it thanks to the Thinking Aloud protocol. The preliminary studies regarding its validity presented good results, although a rigorous analysis with a larger number of uses of the methodology is necessary to determine its suitability. The empirical studies carried out following this methodology with the collaboration archaeologists represent a contribution in themselves, given the scarcity of studies of this nature found until now. All these preliminary techniques developed allowed us to produce a solid and tested characterization of cognitive processes in archaeology in the form of a metamodel, which can be used for the construction of similar systems or for use in conceptual terms for any application requiring the definition of cognitive processes in the field.

However, all these techniques adopted and developed only provided the conceptual and design infrastructure to represent and interpret archaeological data sets structure and semantics and archaeologist's cognitive actions when they analyze the archaeological data sets. We also need to define the software assistance that we want to provide to our archaeologists. In Chaps. 4 and 5 we focus on useful visualization patterns for large archaeological data sets. This work is materialized in the form of a hierarchical metamodel which is independent of the field of application and is not exclusive to one type of interface. This gives the mechanism a great degree of flexibility of application for the definition of presentation and interaction mechanisms in interfaces for software assistance systems aimed at the generation of knowledge.

With a broad overview of the existing techniques and new developments, the reader is now prepared for integrating them in a real conceptual solution. We have shown during Chaps. 6 to 10 how a software system for providing assistance to the generation of knowledge in archaeology using software models is conceptualized, designed, prototyped and validated. The software prototype—exposed and validated in Chaps. 11 and 12—which has been implemented also constitutes a unique research tool which illustrates the whole process.

The research process undertaken has a strong interdisciplinary component, not only for its initial approach but also for the need, which was generated during its execution, to use techniques and tools which are characteristic of other disciplines in order to study the chosen field and the possibilities for assistance within it. For this reason, we believe that, apart from the specific contributions listed, the main contribution of the research presented here is the bi-directional relationship which has been established between both disciplines, given that (1) research has not only been carried out in software engineering, producing a framework supporting assistance in a little-explored field of application and integrating, both methodologically and technically, discourse analysis into the process of software development, but also (2) the research frontiers in archaeology have also been advanced in terms of the improvement of its large data sets analysis and stratigraphic visualization systems, the application of Thinking Aloud techniques for the study of knowledge generation in the field and the possibility of analyzing its textual sources and formally modeling the results.

Yet, as we mentioned several times in this work, there is still a long way to go on software-assisted knowledge generation for archaeological data. Next section takes a look at possible connections and research challenges in the area.

Looking Beyond: The Future of Software-Assisted Knowledge Generation Based on Archaeological Data

As this book points out over the course of several chapters, software engineering and archaeology presents a fluent relationship in last years. Below, we shall outline which how the reader can connect the work presented here with emerging areas in both disciplines and what are the implications fruit of the proposals put forward in this book. They are listed from lowest to highest in terms of scope and implication: from the most immediate actions for the improvement of the solutions proposed here, via the implications and areas of interest which are open in the short to medium term, to possible complex lines of research which may be developed in the future.

As far as areas for improvement in the techniques proposed here are concerned, we believe that, in the short term, a validation of the models proposed as part of the conceptual framework through a broad range of case studies is necessary, especially in the limits of the archaeological studies or data sets. Although the subject model has been tested with real archaeological data and the rest of the models which make up the framework are independent in nature from the field of application due to their abstract definition, the cognitive processes and software presentation and interaction models have only been tested in case studies belonging historical and archaeological contexts (due to our main goal was to cover them). Complex related areas, with data commonly integrated into archaeological data sets, such as anthropological and artistic data may offer case studies which put the models to the test in terms of conceptual representation and suitability of the software assistance offered to their professionals. In the same way, the exhaustive validation of these models would enable us to detect other cognitive processes in archaeology, which may be assisted, and to find out whether there are other software presentation and inter-action patterns which should be incorporated into the proposed conceptual framework.

In addition to the work of improving the proposed solutions, the conceptual framework presented here opens up other possibilities in the medium term regarding studies in software assistance. We consider it especially relevant to promote:

– Studies of the connection between cognitive processes and solutions in terms of software interaction, including the exploration, in software assistance spheres, of other proposals made in terms of the integration of cognitive processes in software engineering, such as approaches based on Intention Mining [150, 181, 235], and of established graphic interface representation language proposals [2]. These approaches may enrich expressiveness and connect the solution presented in this book with existing studies, expanding them to applications related to software assistance, in which they have not yet been applied.
– Studies on how software analysts working on heritage data, archaeologists and related professionals use this conceptual framework in order to define software assistance (which cognitive processes they choose to assist, which presentation and interaction hierarchy pattern they employ, etc.).
– In relation to the studies of use mentioned above, we believe it is necessary to define new metrics in order to evaluate the level of assistance obtained in knowledge generation in any area or field and that the empirical work presented here should offer a solid foundation in order to continue working along these lines.
– Developments complementing those presented here (basically focused on model-directed interoperability) in order to incorporate model-directed software paradigms into the conceptual framework presented. This will provide the proposed solution with a greater degree of dynamism and will enable semi-automatic solutions for instantiation of the conceptual framework to be explored, along with evolutionary models for each of the aspects dealt with in the framework.
– Studies on the connection between the work carried out and the application of discourse analysis techniques in software engineering, especially in the requirements phase, in which their connection with existing notations and approaches in requirements engineering has already been explored. The aim of these studies is to attempt to promote these approaches with a linguistic and philosophical vision and to close the gap which exists between the specifications of textual requirements and the closest requirements models, such as Mind Maps [37, 74], Use Cases [139], and goal-oriented notations such as i* [41, 88, 111, 289] or KAOS [63, 270].

Finally, we believe that it is important to highlight the fact that this book is set in a research context in which co-researcher and transdisciplinary approaches such as this one between software engineering and archaeology and related heritage disciplines have been gaining in importance over recent years. A good example of this are the studies grouped together under the name of Digital Humanities [257]. The current relevance of these approaches in the research community [93, 95] allows our research to be a starting point, not a final destination, in research on this relationship, opening up new lines, and sub-lines, of research in the long term, the most important aspects of which are:

- Lines of research relating to the application of Thinking Aloud Protocols (TAP) as a technique for the characterization, definition and extraction of cognitive processes in Humanities areas.
- The continuation of the line of research initiated regarding the integration of the methodology and of the discourse analysis language presented here in any research process which requires the analysis, structuring and semantic extraction of textual sources written in a free style. Some particularly interesting aspects here are:

  • The application of the methodology and the language presented here in numerous cases within the process of software development. Initially, the possibility of incorporating them into processes of analysis and requirements structuring has been explored with good results [176]. The continuation of this line of research may offer good results in other similar processes in software engineering, such as automatic translation, the indexing of textual sources [35], annotation systems, etc.
  • The application of the methodology and the language presented here in numerous cases within the analysis of scientific valuations and methodological literature in heritage disciplines, encapsulating what knowledge has been generated based on particular data about our past and allowing us to extract methodological entities from the texts. These analyses would allow us to trace the process of knowledge generation about our past more clearly. This approaches could allow us to continue currently works related to different uses of textual information in digital humanities [55, 94, 214, 218, 252, 253].

- The continuation of the line of research initiated regarding software assistance, expanding the types of assistance offered, placing particular emphasis on software assistance via knowledge extraction techniques, which was initially proposed in the research planning stages, but was only explored during the case study [75]. This line of research aims to study the relation between the discourse analysis language presented here and the possibilities for data-mining and text-mining associated with it. This will allow us to incorporate a type of software assistance into the conceptual framework which does not only work by way of adapted visualizations, but also via recommendations about which techniques of "Textual and Data Mining" (TDM) may offer better results according to the type of documental corpus being analyzed. As far as this line of research is concerned, the case study carried out in archaeology, offers new possibilities for the application of natural language processing techniques for the extraction of heritage information on both inferential and methodological levels of discourse.

To sum up, through our work and other similar studies, it is hoped that the relationship between software engineering and archaeology or even nearly heritage disciplines will continue to be transformed, thereby favoring the bidirectional application of research results between the two disciplines. In this future scenario, archaeologists will not simply be a user of software tools which he/she applies in a specific process within the chain of knowledge generation. Rather, as experts, they will integrate parts of the software engineering corpus into their work methodology in the generation of knowledge.

References

1. *Coursera.* Accessed on 26/04/2016. https://www.coursera.org.
2. Brambilla, M., Fraternali, P., et al. *The interaction flow modeling language (ifml)*, version 1.0. Technical report, Object Management Group (OMG) (2014), http://www.ifml.org
3. Abadie, A., A. Diamond, and J. Hainmueller. 2012. Synthetic control methods for comparative case studies: Estimating the effect of California's tobacco control program. *Journal of the American Statistical Association* 105, no. 490 (2010): 493–505. doi:10.1198/jasa.2009. ap08746. Publisher: American Statistical Association.
4. Abdulhadi, S. 2007. i* Guide version 3.0.
5. Abejón Peña, T. 2000. Un instrumento versátil e idóneo para la interconexión de sistemas de información: Tesauro de Patrimonio Histórico Andaluz. *Boletín del Instituto Andaluz del Patrimonio Histórico* 31: 134–141.
6. Aboal-Fernández, R., C. Cancela Cereijo, F. Carrera Ramírez, V. Castro Hierro, P. Mañana-Borrazás, and Y. Porto Tenreiro. 2012. Intervencións de conservación e recuperación no xacemento de Forno dos Mouros (Toques, A Coruña). *Intervenciones de conservación y recuperación en el yacimiento de Forno dos Mouros (Toques, A Coruña).*
7. Ackoff, R.L. 1988. From data to wisdom. *Journal of Applied Systems Analysis* 16: 3–9.
8. Ager, A., and A. Strang. 2008. Understanding integration: A conceptual framework. *Journal of Refugee Studies* 21 (2): 166–191.
9. Akyel, A., and S. Kamisli. 1996. Composing in First and Second Languages: Possible Effects of EFL Writing Instruction.
10. Alexander, C. 1979. *The timeless way of building.* New York: Oxford University Press.
11. Altintas, I., C. Berkley, E. Jaeger, M. Jones, B. Ludascher, and S. Mock. 2004. Kepler: An extensible system for design and execution of scientific workflows. In *Proceedings 16th International Conference on Scientific and Statistical Database Management, 2004.* IEEE, 423–424.
12. Amar, R., J. Eagan, and J. Stasko. 2005. Low-level components of analytic activity in information visualization. In *INFOVIS 2005. IEEE Symposium on Information Visualization, 2005.* IEEE. 111–117.
13. Andrews, K. 2006. Evaluating information visualisations. In *Proceedings of the 2006 AVI Workshop on Beyond Time and Errors: Novel Evaluation Methods for Information Visualization.* ACM. Venice, Italy, 1–5.
14. Andrienko, N., G. Andrienko, and P. Gatalsky. 2003. Exploratory spatio-temporal visualization: an analytical review. *Journal of Visual Languages & Computing* 14 (6): 503–541.
15. Arah, O.A., N.S. Klazinga, D. Delnoij, A. Ten Asbroek, and T. Custers. 2003. Conceptual frameworks for health systems performance: A quest for effectiveness, quality, and improvement. *International Journal for Quality in Health Care* 15 (5): 377–398.

© Springer International Publishing AG 2018
P. Martin-Rodilla, *Digging into Software Knowledge Generation in Cultural Heritage*, Modeling and Optimization in Science and Technologies 11,
https://doi.org/10.1007/978-3-319-69188-6

16. Araujo, J., D. Zowghi, and A. Moreira. 2007. An evolutionary model of requirements correctness with early aspects. In *Ninth International Workshop on Principles of Software Evolution: In Conjunction with the 6th ESEC/FSE Joint Meeting*. ACM. Dubrovnik, Croatia, 67–70.

17. ARIADNE. 2013. Accessed on 26/04/2016. http://ariadne-infrastructure.eu/.

18. Ashcraft, M.H. 1987. Children's knowledge of simple arithmetic: A developmental model and simulation. In *Formal methods in developmental psychology*, 302–338. New York: Springer.

19. Asher, N. 2012. Implicatures in discourse. *Discourse and Grammar: From Sentence Types to Lexical Categories* 112: 11.

20. Asher, N., and S. Pogodalla. 2011. SDRT and continuation semantics. In *New Frontiers in Artificial Intelligence*, 3–15. Heidelberg: Springer.

21. Averbuch, M., I. Cruzgif, W. Lucas, and M. Radzyminski. 2004. As You Like It: Tailorable Information Visualization. Technical Report. (Database Visualization Research Group: Tufts University, 2004).

22. Baeza-Yates, R., and B. Ribeiro-Neto. 1999. *Modern information retrieval*, vol. 463. New York: ACM Press.

23. Baeza-Yates, R., and A. Tiberi. 2007. Extracting semantic relations from query logs. In *Proceedings of the 13th ACM SIGKDD international conference on Knowledge discovery and data mining*, 76–85. ACM.

24. Bamboo. 2008–2012. *Bamboo project*. Accessed on 26/04/2016. http://www.projectbamboo.org/.

25. Beale, J.M., and F.C. Keil. 1995. Categorical effects in the perception of faces. *Cognition* 57 (3): 217–239.

26. Beebe, N.L., and J.G. Clark. 2007. Digital forensic text string searching: Improving information retrieval effectiveness by thematically clustering search results. *Digital Investigation* 4: 49–54.

27. Bellinger, G. 1997. *Knowledge Management—Emerging Perspectives*. http://www.systems-thinking.org/kmgmt/kmgmt.htm.

28. Benardou, A., P. Constantopoulos, C. Dallas, and D. Gavrilis. 2010. Understanding the information requirements of arts and humanities scholarship. *International Journal of Digital Curation* 5 (1): 18–33.

29. Bézivin, J., F. Jouault, and P. Valduriez. 2004. On the need for megamodels. In *Proceedings of the OOPSLA/GPCE: Best Practices for Model-Driven Software Development Workshop, 19th Annual ACM Conference on Object-Oriented Programming, Systems, Languages, and Applications*.

30. Binding, C., K. May, and D. Tudhope. 2008. Semantic interoperability in archaeological datasets: Data mapping and extraction via the CIDOC CRM. In *Research and Advanced Technology for Digital Libraries*, 280–290. Berlin: Springer.

31. Blanke, T., and M. Hedges. 2013. Scholarly primitives: Building institutional infrastructure for humanities e-Science. *Future Generation Computer Systems* 29 (2): 654–661.

32. Booch, G. 1995. *Object Solutions: Managing the Object-Oriented Project*. Addison Wesley Longman Publishing Co., Inc.

33. Bostock, M., V. Ogievetsky, and J. Heer. 2011. D^3 data-driven documents. *IEEE Transactions on Visualization and Computer Graphics* 17 (12): 2301–2309.

34. Brambilla, M., P. Fraternali, and E. Molteni. 2009. A tool for model-driven design of rich internet applications based on AJAX. *Handbook of Research on Web* 2 (3.0): 96–118.

35. Brisaboa, N.R., A. Fariña, G. Navarro, A.S. Places, and E. Rodríguez. 2008. Self-indexing natural language. In *String Processing and Information Retrieval*, 121–132. Berlin: Springer.

36. Brooks, R. 1977. Towards a theory of the cognitive processes in computer programming. *International Journal of Man-Machine Studies* 9 (6): 737–751.

37. Buzan, T., and B. Buzan. 2002. *How to Mind Map*. London: Thorsons.

38. Cairo, A. 2012. *The Functional Art: An Introduction to Information Graphics and Visualization*. Berkeley: New Riders.

39. Campbell, J.I., and J.M. Clark. 1989. Time course of error priming in number-fact retrieval: Evidence for excitatory and inhibitory mechanisms. *Journal of Experimental Psychology: Learning, Memory, and Cognition* 15 (5): 920.

40. Card, S. (2007). Information visualization. In *The Human-Computer Interaction Handbook: Fundamentals, Evolving Technologies, and Emerging Applications*, ed. A. Sears and J.A. Jacko. Mahwah: Lawrence Erlbaum Assoc Inc.

41. Cares, C., X. Franch, A. Perini, and A. Susi. 2011. Towards interoperability of i* models using iStarML. *Computer Standards & Interfaces* 33 (1): 69–79.

42. Carlson, L., D. Marcu, and M.E. Okurowski. 2003. Building a discourse-tagged corpus in the framework of rhetorical structure theory. In *Current and New Directions in Discourse And Dialogue*, 85–112. Berlin: Springer.

43. Carpendale, S. 2008. Evaluating information visualizations. In *Information Visualization*, ed. A. Kerren, et al. Lecture Notes in Computer Science, 19–45. Berlin, Heidelberg: Springer.

44. Carpenter, S.A. 2009. *New Methodology for Measuring Information, Knowledge, and Understanding Versus Complexity in Hierarchical Decision Support Models*. ProQuest.

45. Carpenter, S.A., and J. Cannady. 2004. Tool for sharing and assessing models of fusion-based space transportation systems. In *Proceedings of the 40th AIAA/ASME/ SAE/ASEE Joint Propulsion Conference and Exhibit (July 11–14, 2004), Fort Lauderdale, Florida*.

46. Carpenter, S.A., and J. Cannady. 2004. Tool for sharing and assessing models of fusion-based space transportation systems. In *Proceedings of the 40th AIAA/ASME/ SAE/ASEE Joint Propulsion Conference and Exhibit*. July 11–14, 2004. Fort Lauderdale, Florida.

47. Casamayor, A., D. Godoy, and M. Campo. 2010. Identification of non-functional requirements in textual specifications: A semi-supervised learning approach. *Information and Software Technology* 52 (4): 436–445.

48. Clemmensen, T., M. Hertzum, K. Hornbæk, Q. Shi, and P. Yammiyavar. 2008. Cultural cognition in the thinking-aloud method for usability evaluation. *ICIS 2008 Proceedings*, 189.

49. Cleveland, H. 1982. information as a resource. *Futurist* 16 (6): 34–39.

50. COGNISE. 2015. *International Workshop on Cognitive Aspects of Information Systems Engineering (COGNISE)*. Accessed on 26/04/2016. http://is-lin.hevra.haifa.ac.il/cognise/2015/.

51. Cohen, A.M., and W.R. Hersh. 2005. A survey of current work in biomedical text mining. *Briefings in Bioinformatics* 6 (1): 57–71.

52. Cohen, J. 1988. *Statistical Power Analysis for the Behavioral Sciences*, 2nd ed. Hillsdale, NJ.: Erlbaum.

53. Cohen, J. 1992. A power primer. *Psychological Bulletin*. 122: 155–159.

54. Cook, T.D., D.T. Campbell, and A. Day 1979. *Quasi-Experimentation: Design & Analysis Issues for Field Settings*, vol. 351. Boston: Houghton Mifflin.

55. Cordón-Garcia, J., R. Gómez-Díaz, and M.M. Borges. 2015. New publishing and scientific communication ways: electronic edition, digital educational resources. In *Proceedings of the 3rd International Conference on Technological Ecosystems for Enhancing Multiculturality*, 371–373. ACM.

56. Criado-Boado, F., and M.P. Prieto Martínez. 2010. *Reconstruyendo la historia de la comarca de Ulla-Deza (Galicia-Esapaña): Escenarios arqueológicos del pasado*, vol. 41. Editorial CSIC-CSIC Press.

57. Cullum, L. 1998. Encouraging the Reluctant Reader: Using a Think-Aloud Protocol to Discover Strategies for Reading Success. Report for Department of English, Indiana University of Pennsylvania. (ERIC Document Reproduction Service No. ED 420 837)

58. Chang, D.-S., and K.-S. Choi. 2006. Incremental cue phrase learning and bootstrapping method for causality extraction using cue phrase and word pair probabilities. *Information Processing & Management* 42 (3): 662–678.

59. Charters, E. 2003. The use of think-aloud methods in qualitative research an introduction to think-aloud methods. *Brock Education Journal* 12 (2).

60. Chen, J.Q., and S.M. Lee. 2003. An exploratory cognitive DSS for strategic decision making. *Decision support systems.* 36 (2): 147–160.

61. Chen, M., D. Ebert, H. Hagen, R.S. Laramee, R. Van Liere, K.-L. Ma, W. Ribarsky, G. Scheuermann, and D. Silver. 2009. Data, information, and knowledge in visualization. *Computer Graphics and Applications IEEE* 29 (1): 12–19.

62. Chen, P.P.-S. 1976. The entity-relationship model—toward a unified view of data. *ACM Transactions on Database Systems (TODS)* 1 (1): 9–36.

63. Dardenne, A., A. Van Lamsweerde, and S. Fickas. 1993. Goal-directed requirements acquisition. *Science of Computer Programming* 20 (1): 3–50.

64. Darvill, T. 2008. *Concise Oxford Dictionary of Archaeology*, Second edition. Oxford: Oxford University Press.

65. Davenport, T.H. 1993. *Process Innovation: Reengineering Work Through Information Technology*. Boston: Harvard Business Press.

66. Davenport, T.H., J. Harris, and J. Shapiro. 2010. Competing on talent analytics. *Harvard Business Review* 10 (88): 52–58.

67. de Lara, J., and E. Guerra. 2014. Towards the flexible reuse of model transformations: A formal approach based on graph transformation. *Journal of Logical and Algebraic Methods in Programming* 83 (5): 427–458.

68. Del Fabro, M.D., and P. Valduriez. 2007. Semi-automatic model integration using matching transformations and weaving models. In *Proceedings of the 2007 ACM Symposium on Applied Computing*, 963–970. ACM.

69. Doerr, M. 2003. The CIDOC conceptual reference module: An ontological approach to semantic interoperability of metadata. *AI Mag* 24 (3): 75–92.

70. Doerr, M. 2005. The CIDOC CRM, an ontological approach to schema heterogeneity. *Semantic Interoperability and Integration* 4391.

71. Doerr, M., A. Kritsotaki, and K. Boutsika. 2011. Factual argumentation—a core model for assertions making. *Journal on Computing and Cultural Heritage (JOCCH)* 3 (3): 8.

72. Dybå, T., V.B. Kampenes, and D.I. Sjøberg. 2006. A systematic review of statistical power in software engineering experiments. *Information and Software Technology* 48 (8): 745–755.

73. El Hamlaoui, M., S. Ebersold, B. Coulette, M. Nassar, and A. Anwar. 2014. Heterogeneous models matching for consistency management. In *2014 IEEE Eighth International Conference on Research Challenges in Information Science (RCIS)*, 1–12. IEEE.

74. Eppler, M.J. 2006. A comparison between concept maps, mind maps, conceptual diagrams, and visual metaphors as complementary tools for knowledge construction and sharing. *Information Visualization* 5 (3): 202–210.

75. Epure, E.V., P. Martin-Rodilla, C. Hug, R. Deneckère, and C. Salinesi. 2015. Automatic process model discovery from textual methodologies, 19–30.

76. Ericsson, K.A., and H.A. Simon. 1980. Verbal reports as data. *Psychological Review* 87 (3): 215.

77. España, S., A. González, and Ó. Pastor. 2009. Communication Analysis: A Requirements Engineering Method for Information Systems. In *Advanced Information Systems Engineering*, ed. P. Eck, J. Gordijn, and R. Wieringa. Lecture Notes in Computer Science, 530–545. Berlin, Heidelberg: Springer.

78. Europeana. 2008–2015. *Europeana Project*. Accessed on 26/04/2016. http://www.europeana.eu/.

79. Exertier, D., and S. Bonnet. 2014. Arcadia/Capella, a field-proven modeling solution for system and software architecture engineering. France 2014 eclipsecon.

80. Falconer, S.M., and M.-A. Storey. 2007. *A Cognitive Support Framework for Ontology Mapping*. New York: Springer.

81. Faul, F., E. Erdfelder, A. Buchner, and A.-G. Lang, Statistical power analyses using G*Power 3.1: Tests for correlation and regression analyses. *Behavior Research Methods* 41 (4): 1149–1160.

82. Faul, F., E. Erdfelder, A.-G. Lang, and A. Buchner, G*Power 3: A flexible statistical power analysis program for the social, behavioral, and biomedical sciences. *Behavior Research Methods* 39 (2): 175–191.

83. Feagin, J.R., A.M. Orum, and G. Sjoberg. 1991. *A Case for the Case Study*. London: UNC Press Books.

84. Fernández Freire, C., I. del Bosque González, P. Fábrega-Álvarez, A. Fraguas Bravo, C. Parcero-Oubiña, E. Pérez Asensio, A. Uriarte González, and J.M. Vicent García. 2012. Cultural heritage application schema: A SDI framework within the protected sites INSPIRE spatial data theme. In *Computer Applications and Quantitative Methods in Archaeology (CAA) 2012*. Southampton, UK.

85. Fontana, A., and J.H. Frey. 2000. The interview: From structured questions to negotiated text. *Handbook of Qualitative Research* 2 (6): 645–672.

86. Foster, S.J., J.D. Hoge, and R.H. Rosch. 1999. Thinking aloud about history: Children's and adolescents' responses to historical photographs. *Theory & Research in Social Education*. 27 (2): 179–214.

87. Eclipse Foundation. 2014. *EMF Eclipse Project*. Accessed on 04/11/2014. http://www.eclipse.org/modeling/emf/.

88. Fuxman, A., L. Liu, J. Mylopoulos, M. Pistore, M. Roveri, and P. Traverso. 2004. Specifying and analyzing early requirements in tropos. *Requirements Engineering* 9 (2): 132–150.

89. Gamma, E., R. Helm, R. Johnson, and J. Vlissides. 1995. *Design Patterns: Elements of Reusable Object-Oriented Software*, 395. Addison-Wesley: Longman Publishing Co., Inc.

90. Gardin, J.-C. 2002. Archaeological discourse, conceptual modelling and digitalisation: An interim report of the logicist program, 5–11. In *CAA*.

91. Gibson, B. 1997. Talking the test: Using verbal report data in looking at the processing of cloze tasks. *Edinburgh Working Papers in Applied Linguistics* 8: 54–62.

92. Goble, C., R. Stevens, D. Hull, K. Wolstencroft, and R. Lopez. 2008. Data curation+ process curation = data integration+ science. *Briefings in Bioinformatics* 9 (6): 506–517.

93. González-Blanco, E. 2007. The state of humanities in Spain. *International Journal of the Humanities* 5 (1).

94. González-Blanco, E., and J.L. Rodríguez. 2014. ReMetCa: A proposal for integrating RDBMS and TEI-Verse. *Journal of the Text Encoding Initiative* 8.

95. Gonzalez-Blanco Garcia, E. 2013. Digital Humanities news and an example of poetic assembly on the net. *CUADERNOS HISPANOAMERICANOS* 761: 53–67.

96. Gonzalez-Perez, C. 2010–2013. *ConML*. Accessed on 26/04/2016. http://www.conml.org/.

97. Gonzalez-Perez, C. 2012. A conceptual modelling language for the humanities and social sciences. In *RCIS'12: Sixth International Conference on Research Challenges in Information Science*, 1–6. Xplore. Valencia, Spain.

98. Gonzalez-Perez, C. 2012. Typeless information modelling to avoid category bias in archaeological descriptions. In *Thinking Beyond the Tool: Archaeological Computing & the Interpretive Process*, ed. P.M.F.A. Chrysanthi, and C. Papadopoulos. Oxford, UK: Archaeopress.

99. González-Pérez, C. 2010–2013. *CHARM: Cultural Heritage Abstract Reference Model*. Accessed on 26/04/2016. http://www.charminfo.org/.

100. Gonzalez-Perez, C., and B. Henderson-Sellers. 2005. A representation-theoretical analysis of the OMG modelling suite, 252–262 In *SoMeT*.

101. Gonzalez-Perez, C., and B. Henderson-Sellers. 2006. An ontology for software development methodologies and endeavours. In *Ontologies for Software Engineering and Software Technology*, 123–151. Berlin: Springer.

102. Gonzalez-Perez, C., and B. Henderson-Sellers. 2006. A powertype-based metamodelling framework. *Software & Systems Modeling* 5 (1): 72–90.

103. González-Pérez, C., and C. Hug. 2013. Crafting archaeological methodologies: Suggesting method engineering for the humanities and social sciences. In *Computer Applications and Quantitative Methods in Archaeology (CAA) 2012*. Southampton, UK.

104. González-Pérez, C., and P. Martin-Rodilla. 2014. Integration of Archaeological Datasets Through the Gradual Refinement of Models. In *Proceedings of the 42nd Annual Conference on Computer Applications and Quantitative Methods in Archaeology, CAA 2014*. Paris, France.

105. Gonzalez-Perez, C., P. Martin-Rodilla, C. Parcero-Oubiña, P. Fábrega-Álvarez, and A. Güimil-Fariña. 2012. Extending an Abstract Reference Model for Transdisciplinary Work in Cultural Heritage, 190–201.

106. González-Pérez, C., and C. Parcero-Oubiña. 2011. A Conceptual Model for Cultural Heritage Definition and Motivation. In *Revive the Past. Computer Applications and Quantitative Methods in Archaeology (CAA)*, 234–244. *Proceedings of the 39th International Conference*, ed. M. Zhou, et al. Amsterdam: Pallas Publications.

107. Graesser, A.C., D.S. McNamara, and M.M. Louwerse. 2010. 2 Methods of automated text analysis. *Handbook of Reading Research* 4: 34.

108. Graumlich, J.F., N.L. Novotny, G. Stephen Nace, H. Kaushal, W. Ibrahim-Ali, S. Theivanayagam, L. William Scheibel, and J.C. Aldag. 2009. Patient readmissions, emergency visits, and adverse events after software-assisted discharge from hospital: Cluster randomized trial. *Journal of Hospital Medicine* 4 (7): E11–E19.

109. Grissom, R.J., and J.J. Kim. 2005. *Effect Sizes for Research. A Broad Practical Approach*. Mahwah, NJ: Lawrence Erlbaum.

110. Groen, G.J., and J.M. Parkman. 1972. A chronometric analysis of simple addition. *Psychological Review* 79 (4): 329.

111. Guizzardi, R.S., and A. Perini. 2005. Analyzing requirements of knowledge management systems with the support of agent organizations. *Journal of the Brazilian Computer Society* 11 (1): 51–62.

112. Hadar, I. 2013. When intuition and logic clash: The case of the object-oriented paradigm. *Science of Computer Programming* 78 (9): 1407–1426.

113. Harmain, H.M., and R. Gaizauskas. 2000. CM-Builder: An automated NL-based CASE tool. In *The Fifteenth IEEE International Conference on Automated Software Engineering, 2000. Proceedings ASE 2000*, 45–53. IEEE.

114. Harris, E. 1979. *Principles of Archaeological Stratigraphy*. London & New York: Academic Press.

115. Harris, Z.S. 1952. Discourse analysis: A sample text. *Language* 28: 474–494.

116. Hayes, J.H., A. Dekhtyar, and J. Osborne. 2003. Improving requirements tracing via information retrieval. In *11th IEEE International Requirements Engineering Conference, 2003. Proceedings*, 138–147. IEEE.

117. Henderson-Sellers, B., and C. Gonzalez-Perez. 2005. The rationale of powertype-based metamodelling to underpin software development methodologies. In *Proceedings of the 2nd Asia-Pacific conference on Conceptual modelling-Volume 43*, 7–16. Australian Computer Society, Inc.

118. Henderson-Sellers, B., and C. Gonzalez-Perez. 2010. Granularity in conceptual modelling: Application to metamodels. In *Conceptual Modeling–ER 2010*, 219–232. Heidelberg: Springer.

119. Herbst, H. 1997. *Business Rule-Oriented Conceptual Modeling*. Springer Science & Business Media.

120. Hobbs, J.R. 1985. *On the Coherence and Structure of Discourse*. CSLI.

121. Holibaugh, R.R., J.M. Perry, and L. Sun. 1988. Phase I Testbed Description: Requirements and Selection Guidelines. Techical Report CMU/SEI-88-TR-13, Software Engineering Institute, Carnegie Mellon University.
122. Holzinger, A. 2005. Usability engineering methods for software developers. *Communications of the ACM* 48 (1): 71–74.
123. IBM, Released 2013. IBM SPSS Statistics for Windows, Version 23.0. IBM Corp. Armonk, NY.
124. Apple Inc. 2015. *iOS Developer Library*. Accessed on 30/03/2016. https://developer.apple.com/library/ios/navigation/.
125. Apple Inc. 2015. *iOS SDK Release Notes for iOS 8.4*. Accessed on 30/03/2016. https://developer.apple.com/library/ios/releasenotes/General/RN-iOSSDK-8.4/index.html#//apple_ref/doc/uid/TP40015246.
126. Apple Inc. 2016. *Core Data Programming Guide*. Accessed on 30/03/2016. https://developer.apple.com/library/ios/documentation/Cocoa/Conceptual/CoreData/index.html.
127. Incipit. 2014. *Cultural Heritage Abstract Reference Model, CHARM 0.8*.
128. Incipit. 2014. *CHARM Extension Guidelines Version 1.0.1* Accessed on 30/03/2016. http://www.charminfo.org/Resources/Technical.aspx.
129. Incipit. 2015. *ConML Technical Specification. ConML 1.4.4*. http://www.conml.org/Resources_TechSpec.aspx.
130. Incipit. 2015. *Cultural Heritage Abstract Reference Model, CHARM 0.9.0.2*. Accessed on 30/03/2016. http://www.charminfo.org/Resources/Default.aspx.
131. Instituto Español de Estadística, I. 2009. *Encuesta sobre Recursos Humanos en Ciencia y Tecnología. Año 2009. Características del Doctorado: Porcentaje de doctores por campo de doctorado y sexo*. Accessed on 30/03/2016. http://www.ine.es/jaxi/tabla.do?path=/t14/p225/a2009/l0/&file=02001.px&type=pcaxis&L=0.
132. Instituto Nacional de Estadística, I. 2011. *Estadística de la Enseñanza Universitaria en España. Curso 2010–2011. Resúmenes Generales: Personal docente de los centros propios de las Universidades Públicas por Área de conocimiento, sexo y Categoría*. Accessed on 30/03/2016. http://www.ine.es/jaxi/tabla.do?type=pcaxis&path=/t13/p405/a2010-2011/l0/&file=02006.px.
133. ISO/IEC. 2001. *ISO/IEC 9126-1 Software Engineering—Product Quality-1: Quality Model*.
134. ISO/IEC. 2006. Topics Maps. ISO/IEC 13250/2006.
135. ISO/IEC. 2009. ISO/IEC 20926:2009 Software and systems engineering—Software measurement—IFPUG functional size measurement method.
136. ISO/IEC. 2010. *ISO/IEC/IEEE 24765: 2010 Systems and Software Engineering–Vocabulary*.
137. ISO/IEC. 2011. *ISO/IEC 25010:2011 Systems and software engineering—Systems and software Quality Requirements and Evaluation (SQuaRE)—System and software quality models*.
138. ISO/IEC. 2011. ISO/IEC 25040:2011 Systems and software engineering—Systems and software Quality Requirements and Evaluation (SQuaRE)—Evaluation process.
139. ISO/IEC. 2012. Information technology—Object Management Group Unified Modeling Language (OMG UML) Part 1: Infrastructure. ISO/IEC 19505-1:2012.
140. ISO/IEC. 2014. ISO 24744: 2014 Software Engineering—Metamodel for Development Methodologies.
141. Jabareen, Y. 2009. Building a conceptual framework: Philosophy, definitions, and procedure. *International Journal of Qualitative Methods* 8 (4): 49–62.
142. Jacobson, I., G. Booch, and J. Rumbaugh, 1999. *The unified software development process*, vol. 1. Reading: Addison-Wesley.
143. Jaworski, B.J., V. Stathakopoulos, and H.S. Krishnan. 1993. Control combinations in marketing: Conceptual framework and empirical evidence. *The Journal of Marketing* 57–69.

144. Jensen, L.J., J. Saric, and P. Bork. 2006. Literature mining for the biologist: From information retrieval to biological discovery. *Nature Reviews Genetics* 7 (2): 119–129.

145. Jørgensen, A.H. 1990. Thinking-aloud in user interface design: A method promoting cognitive ergonomics. *Ergonomics* 33 (4): 501–507.

146. Juristo, N., J.L. Morant, and A.M. Moreno. 1999. A formal approach for generating OO specifications from natural language. *Journal of Systems and Software* 48 (2): 139–153.

147. Juristo, N., and A.M. Moreno. 2001. *Basics of Software Engineering Experimentation.* Germany: Springer.

148. Keim, D., G. Andrienko, J.-D. Fekete, C. Görg, J. Kohlhammer, and G. Melançon. 2008. *Visual Analytics: Definition, Process, and Challenges.* New York: Springer.

149. Kelly, S., and J.P. Tolvanen. 2008. *Domain-Specific Modeling: Enabling Full Code Generation.* New York: Wiley.

150. Khodabandelou, G., C. Hug, R. Deneckere, and C. Salinesi. 2013. Process mining versus intention mining. In *Enterprise, Business-Process and Information Systems Modeling,* 466–480. Berlin: Springer.

151. King, E.G., and R.J. Hobbs. 2006. Identifying linkages among conceptual models of ecosystem degradation and restoration: Towards an integrative framework. *Restoration Ecology* 14 (3): 369–378.

152. Kitchenham, B., S.L. Pfleeger, L.M. Pickard, P.W. Jones, D.C. Hoaglin, K. El Emam, and J. Rosenberg. 2002. Preliminary guidelines for empirical research in software engineering. *IEEE Transactions on Software Engineering* 28 (8): 721–734.

153. Knott, A., and T. Sanders. 1998. The classification of coherence relations and their linguistic markers: An exploration of two languages. *Journal of Pragmatics* 30 (2): 135–175.

154. Kof, L. 2007. Scenarios: Identifying missing objects and actions by means of computational linguistics. In *Requirements Engineering Conference. RE'07. 15th IEEE International,* 121–130. IEEE.

155. Kosara, R., C.G. Healey, V. Interrante, D.H. Laidlaw, and C. Ware. 2003. User studies: Why, how, and when? *IEEE Computer Graphics and Applications* 4: 20–25.

156. Kovacic, I. 2000. Thinking-aloud protocol-interview-text analysis. *Benjamins Translation Library* 37: 97–110.

157. Krahmer, E., and N. Ummelen. 2004. Thinking about thinking aloud: A comparison of two verbal protocols for usability testing. *IEEE Transactions on Professional Communication* 47 (2): 105–117.

158. Lacson, R.C., R. Barzilay, and W.J. Long. 2006. Automatic analysis of medical dialogue in the home hemodialysis domain: Structure induction and summarization. *Journal of Biomedical Informatics* 39 (5): 541–555.

159. Lantes Suárez, Ó., A. Martínez Cortizas, and M.P. Prieto-Martínez. 2008. O campaneiforme cordado de Forno dos Mouros (Toques, A Coruña). *The corded bell beaker of Forno dos Mouros (Toques, A Coruña).*

160. Lapata, M., and A. Lascarides. 2006. Learning sentence-internal temporal relations. *J. Artif. Intell. Res. (JAIR)* 27: 85–117.

161. Leamy, M., V. Bird, C. Le Boutillier, J. Williams, and M. Slade. 2011. Conceptual framework for personal recovery in mental health: Systematic review and narrative synthesis. *The British Journal of Psychiatry* 199 (6): 445–452.

162. Lewis, C. 1982. *Using the "Thinking-Aloud" Method in Cognitive Interface Design.* IBM TJ Watson Research Center.

163. Likert, R. 1932. A technique for the measurement of attitudes. *Archives of psychology* 140: 5.

164. Limbourg, Q., J. Vanderdonckt, B. Michotte, L. Bouillon, and V. López-Jaquero. 2005. USIXML: A language supporting multi-path development of user interfaces. In *Engineering Human Computer Interaction and Interactive Systems,* 200–220. Berlin: Springer.

165. Lindvall, M., I. Rus, P. Donzelli, A. Memon, M. Zelkowitz, A. Betin-Can, T. Bultan, C. Ackermann, B. Anders, and S. Asgari. 2007. Experimenting with software testbeds for evaluating new technologies. *Empirical Software Engineering* 12 (4): 417–444.

166. Lindvall, M., I. Rus, F. Shull, M. Zelkowitz, P. Donzelli, A. Memon, V. Basili, P. Costa, R. Tvedt, and L. Hochstein. 2005. An evolutionary testbed for software technology evaluation. *Innovations in Systems and Software Engineering* 1 (1): 3–11.

167. Liskov, B.H., and J.M. Wing. 1994. A behavioral notion of subtyping. *ACM Transactions on Programming Languages and Systems (TOPLAS)* 16 (6): 1811–1841.

168. Lortal, G., S. Dhouib, and S. Gérard. 2011. Integrating ontological domain knowledge into a robotic DSL. In *Models in Software Engineering*, 401–414. Heidelberg: Springer.

169. Mañana-Borrazás, P. 2003. Vida y muerte de los Megalitos. ¿Se abandonan los Túmulos?.

170. Marr, D., and A. Vision. 1982. *A Computational Investigation into the Human Representation and Processing of Visual Information*. San Francisco: WH Freeman and Company.

171. Martin-Rodilla, P. 2013. Empirical approaches to the analysis of archaeological discourse. In *Across Space and Time. Papers from the 41st Annual Conference of Computer Applications and Quantitative Methods in Archaeology (CAA), Amsterdam University Press, Amsterdam*. Perth, 25–28 March 2013.

172. Martin-Rodilla, P. 2013. Knowledge-Assisted Visualization in the Cultural Heritage Domain-Case Studies, Needs and Reflections, 546–549 In *GRAPP/IVAPP*.

173. Martín-Rodilla, P. 2012. The role of software in cultural heritage issues: Types, user needs and design guidelines based on principles of interaction. In *RCIS'12: Sixth International Conference on Research Challenges in Information Science*, 1–2. IEEE XPlore. Valencia, Spain.

174. Martín-Rodilla, P. 2013. Software-assisted knowledge generation in the archaeological domain: A conceptual framework. In *25th International Conference on Advanced Information Systems Engineering (CAiSE 2013): Doctoral Consortium*, ed. B.W.E. Marta Indulska. Valencia, Spain.

175. Martín-Rodilla, P., G. Giachetti, and C. Gonzalez-Perez. 2015. Achieving software-assisted knowledge generation through model-driven interoperability. In *Jornadas de Ingeniería del Software y Bases de Datos (JISBD)*. September 2015. Actas de las XX Jornadas de Ingeniería del Software y Bases de Datos (JISBD 2015). Santander (Spain).

176. Martin-Rodilla, P., and C. Gonzalez-Perez. 2014. An ISO/IEC 24744-derived modelling language for discourse analysis, 1–10.

177. Martín-Rodilla, P., C. Gonzalez-Perez, and P. Mañana-Borrazas. 2015. A conceptual and visual proposal to decouple material and interpretive information about stratigraphic data. In *43rd Annual Conference on Computer Applications and Quantitative Methods in Archaeology (CAA 2015)*. 30/03–3/04/2015. Siena (Italy).

178. Martín-Rodilla, P., C. González-Pérez, J.I. Panach, and Ó. Pastor. An experiment on accuracy, efficiency, productivity and researchers? Satisfaction in digital humanities data analysis: Dataset appendix.

179. Martin-Rodilla, P., J.I. Panach, and O. Pastor. 2014. User interface design guidelines for rich applications in the context of cultural heritage data. In *2014 IEEE Eighth International Conference on Research Challenges in Information Science (RCIS)*, 1–10. IEEE.

180. Mc Kevitt, P., D. Partridge, and Y. Wilks. 1999. Why machines should analyse intention in natural language dialogue. *International Journal of Human-Computer Studies* 51 (5): 947–989.

181. Mei, T., X.-S. Hua, H.-Q. Zhou, and S. Li. 2007. Modeling and mining of users' capture intention for home videos. *IEEE Transactions on Multimedia* 9 (1): 66–77.

182. Menck, V. 2004. Proceedings of Workshop on Intelligent Technologies for Software Engineering (WITSE04, Sept 21, 2004, part of ASE 2004). Deriving behavior specifications from textual use cases. In *Workshop on Intelligent Technologies for Software Engineering (WITSE04, Sept 21, 2004, part of ASE 2004)*, 331–341. Linz, Austria.

183. Menéndez Fernández, M., A. Jimeno Martínez, and V.M. Fernández Martínez. 1997. *Diccionario de prehistoria*. Alianza Editorial.

184. Merriam, S.B. 1998. *Qualitative Research and Case Study Applications in Education. Revised and Expanded from Case Study Research in Education*. ERIC.

185. Microsoft. 2013. Microsoft Excel version 2013. Microsoft Corporation.

186. Mich, L. 1996. NL-LOOPS : From natural language to object oriented requirements using the natural language processing system LOLITA. *Natural Language Engineering* 2: 161–187.

187. Moens, M.-F., E. Boiy, R.M. Palau, and C. Reed. 2007. Automatic Detection of Arguments in Legal Texts, 225–230. ACM.

188. Molina Moreno, P.J. 2003. Especificación del interfaz de usuario: De los requisitos a la generación automática. Universitat Politècnica de València.

189. Molina, P., S. Meliá, and O. Pastor. 2002. JUST-UI: A user interface specification model. In *Computer-Aided Design of User Interfaces III*, ed. C. Kolski and J. Vanderdonckt, 63–74. Springer Netherlands.

190. Moody, D.L. 2003. The method evaluation model: A theoretical model for validating information systems design methods. *ECIS 2003 Proceedings*, 79.

191. Moore, J.D., and P. Wiemer-Hastings. 2003. Discourse in computational linguistics and artificial intelligence. In *Handbook of Discourse Processes*, 439–487.

192. Moreno, A.M. 1997. Object-oriented analysis from textual specifications. In *Ninth International Conference on Software Engineering and Knowledge Engineering, Madrid, Spain (June 1997)*.

193. Munzner, T. 2009. A nested model for visualization design and validation. *IEEE Transactions on Visualization and Computer Graphics* 15 (6): 921–928.

194. Network, Y.D. 2013. *Yahoo Design Pattern Library*. Accessed on 26/04/2016. http://developer.yahoo.com/ypatterns/.

195. Nielsen, J. 1994. *Usability Engineering*. Amsterdam: Elsevier.

196. Ning, Z., J.S. Yi, M.J. Palakal, and A. McDaniel. 2010. OncoViz: A user-centric mining and visualization tool for cancer-related literature. In *Proceedings of the 2010 ACM Symposium on Applied Computing*, 1827–1828. ACM. Sierre, Switzerland.

197. Nunan, D. 1992. *Research Methods in Language Learning*. Cambridge: Cambridge University Press.

198. Nunes, N., and J. Cunha. 2001. Wisdom—A UML based architecture for interactive systems. In *Interactive Systems Design, Specification, and Verification*, ed. P. Palanque and F. Paternò. Lecture Notes in Computer Science, 191–205. Berlin, Heidelberg: Springer.

199. Odell, J.J. 1994. Power types. *Journal of Object-Oriented Programming* 7 (2): 8.

200. Olson, G.M., S.A. Duffy, and R.L. Mack. 1984. Thinking-out-loud as a method for studying real-time comprehension processes. *New Methods in Reading Comprehension Research* 253: 286.

201. OMG. 2007. Software Process Engineering Metamodel Specification. ptc/07-03-03.

202. OMG. 2014. *OMG Meta-Object Facility (MOF) Core Specification, Version 2.4.1. Object Management Group. p. 1*.

203. OMG. 2015. *OMG Unified Modeling Language Specification v.2.5* Object Management Group.

204. Palmer, C.L., L.C. Teffeau, and C.M. Pirmann. 2009. Scholarly information practices in the online environment. *Report Commissioned by OCLC Research. Published online at:* www.oclc.org/programs/publications/reports/2009-02.pdf.

205. Panach, J.I., S. España, O. Dieste, O. Pastor, and N. Juristo. 2015. In search of evidence for model-driven development claims: An experiment on quality, effort, productivity and satisfaction. *Information and Software Technology* 62: 164–186.

206. Parga-Dans, E. 2010. Paper: Innovation and Crisis Era: The Case of Commercial Archaeology as a Knowledge Intensive Business Service.

207. Parga-Dans, E. 2011. *The labour market in the archaeological field: The Spanish contract archaeology*. El mercado de trabajo en el ámbito arqueológico: La arqueología comercial española.

208. Park, J. 2001. *jTME: A Java Topic Map Engineering*. Web Page Available on line at http://xml.coverpages.org/ni2001-03-22-b.html

209. Parsons, J., and Y. Wand. 2013. Extending classification principles from information modeling to other disciplines. *Journal of the Association for Information Systems* 14 (3).

210. Partridge, C. 2005. *Business Objects: Re-Engineering for Re-Use*, 440. 2Rev Ed edition. The BORO Centre.

211. Pastor, O., G. Giachetti, B. Marín, and F. Valverde. 2013. Automating the interoperability of conceptual models in specific development domains. In *Domain Engineering*, ed. I. Reinhartz-Berger, et al., 349–373. Berlin, Heidelberg: Springer.

212. Pastor, O., E. Insfrán, V. Pelechano, J. Romero, and J. Merseguer. 1997. OO-Method: An OO software production environment combining conventional and formal methods. In *Advanced Information Systems Engineering*, ed. A. Olivé and J. Pastor. Lecture Notes in Computer Science, 145–158. Berlin, Heidelberg: Springer.

213. Peffers, K., T. Tuunanen, M.A. Rothenberger, and S. Chatterjee. 2007. A design science research methodology for information systems research. *Journal of Management Information Systems* 24 (3): 45–77.

214. Pereira, J., F. Schmidt, P. Contreras, F. Murtagh, and H. Astudillo. 2010. Clustering and semantics preservation in cultural heritage information spaces. In *Adaptivity, Personalization and Fusion of Heterogeneous Information*, 100–105. LE CENTRE DE HAUTES ETUDES INTERNATIONALES D'INFORMATIQUE DOCUMENTAIRE. Paris, France.

215. Peterson, D. 1996. *Forms of Representation: An Interdisciplinary Theme for Cognitive Science*. Oxford: Intellect Books.

216. Pinggera, J., S. Zugal, M. Furtner, P. Sachse, M. Martini, and B. Weber. 2014. The modeling mind: Behavior patterns in process modeling. In *Enterprise, Business-Process and Information Systems Modeling*, 1–16. Berlin: Springer.

217. Pinggera, J., S. Zugal, and B. Weber. 2010. Investigating the process of process modeling with cheetah experimental platform–tool paper, 13. *ER-POIS 2010*.

218. Pinto Llorente, A.M., M.C. Sánchez Gómez, and F.J. García-Peñalvo. 2016. Assessing the effectiveness of interactive and collaborative resources to improve reading and writing in english. *International Journal of Human Capital and Information Technology Professionals (IJHCITP)* 7 (1): 66–85.

219. Pohl, K. 2010. *Requirements Engineering: Fundamentals, Principles, and Techniques*. Springer Publishing Company, Incorporated.

220. Polanyi, L. 1988. A formal model of the structure of discourse. *Journal of Pragmatics* 12 (5): 601–638.

221. Polarsys.org. 2015. *Capella Software*. Accessed on 30/03/2016. http://www.polarsys.org/capella.

222. Poldrack, R.A. 2006. Can cognitive processes be inferred from neuroimaging data? *Trends in Cognitive Sciences* 10 (2): 59–63.

223. Prasad, R., N. Dinesh, A. Lee, E. Miltsakaki, L. Robaldo, A.K. Joshi, and B.L. Webber. 2008. The Penn Discourse TreeBank 2.0. In *LREC*. Citeseer.

224. Pressley, M., and P. Afflerbach. 1995. *Verbal Protocols of Reading: The Nature of Constructively Responsive Reading*. New York: Routledge.

225. Prieto Martínez, M.P. 2007. Volviendo a un mismo lugar: recipientes y espacios en un monumento megalítico gallego (NW de España). *Revista portuguesa de arqueologia* 10 (2): 101–125.

226. Radatz, J., A. Geraci, and F. Katki. 1990. IEEE standard glossary of software engineering terminology. *IEEE Std* 610121990 (121990): 3.

227. Rasmussen, M.I., J.C. Refsgaard, L. Peng, G. Houen, and P. Højrup. 2011. CrossWork: Software-assisted identification of cross-linked peptides. *Journal of Proteomics* 74 (10): 1871–1883.

228. Ravitch, S.M., and M. Riggan. 2016. *Reason & Rigor: How Conceptual Frameworks Guide Research*. Thousand Oaks: Sage Publications.

229. Reas, C., and B. Fry. 2007. *Processing: A Programming Handbook for Visual Designers and Artists*, vol. 6812. Cambridge: MIT Press.

230. Reddi, V.J., M.S. Gupta, M.D. Smith, W. Gu-Yeon, D. Brooks, and S. Campanoni. 2009. Software-assisted hardware reliability: Abstracting circuit-level challenges to the software stack, 788–793. In *Design Automation Conference, 2009. DAC '09. 46th ACM/IEEE*. 26–31 July 2009.

231. Reitter, D. 2003. Rhetorical analysis with rich-feature support vector models. *Unpublished Master's thesis, University of Potsdam, Potsdam, Germany*.

232. Rishe, N.D., R.I. Athauda, J. Yuan, and S.-C. Chen. 2000. Knowledge management for database interoperability. In *The ISCA 2nd International Conference On Information Reuse And Integration (IRI-2000)*, 23–26. Honolulu, Hawaii.

233. Rolland, C. 2013. Conceptual modeling and natural language analysis. In *Seminal Contributions to Information Systems Engineering*, ed. J. Bubenko, et al., 57–61. Berlin, Heidelberg: Springer.

234. Rolland, C., and C.B. Achour. 1998. Guiding the construction of textual use case specifications. *Data & Knowledge Engineerin*. 25 (1): 125–160.

235. Rolland, C., N. Prakash, and A. Benjamen. 1999. A multi-model view of process modelling. *Requirements Engineering* 4 (4): 169–187.

236. Rolland, C., and C. Proix. 1992. A natural language approach for requirements engineering. In *Advanced Information Systems Engineering*, ed P. Loucopoulos. Lecture Notes in Computer Science, 257–277. Berlin, Heidelberg: Springer.

237. Rosenbloom, P.S. 2010. Towards a conceptual framework for the digital humanities.

238. Russell, M.A. 2008. *Dojo: The Definitive Guide*. Sebastopol, CA (USA): O'Reilly Media, Inc.

239. Sacks, O. 1985. *The Man Who Mistook His Wife for His Hat and Other Clinical Tales*. London: Duckworth.

240. Sánchez-Carretero, C. 2012. *Heritage Regimes and the Camino de Santiago: Gaps and Logics*. Universitätsverlag Göttingen.

241. Santos, J., A. Moreira, J. Araujo, V. Amaral, M. Alférez, and U. Kulesza. 2008. Generating requirements analysis models from textual requirements. In *First International Workshop on Managing Requirements Knowledge. MARK'08*, 32–41. IEEE.

242. Saraiya, P., C. North, and K. Duca. 2005. An insight-based methodology for evaluating bioinformatics visualizations. *IEEE Transactions on Visualization and Computer Graphics* 11 (4): 443–456.

243. Scheithauer, G., and G. Wirtz. 2010. Business modeling for service descriptions: A meta model and a UML profile. In *Proceedings of the Seventh Asia-Pacific Conference on Conceptual Modelling—Volume 110*, 79–88. Brisbane, Australia: Australian Computer Society, Inc.

244. Schreiber, G., and B. Wielinga. 1993. *KADS: A Principled Approach to Knowledge-Based System Development*, vol. 11. London: Academic Press.

245. Schwenk, C.R. 1984. Cognitive simplification processes in strategic decision-making. *Strategic Management Journal* 5 (2): 111–128.

246. Sensalire, M., P. Ogao, and A. Telea. 2009. Evaluation of software visualization tools: Lessons learned. In *5th IEEE International Workshop on Visualizing Software for Understanding and Analysis. VISSOFT 2009*, 19–26. 25–26 Sept 2009.

247. Shields, P.M., and N. Rangarajan. 2013. *A Playbook for Research Methods: Integrating Conceptual Frameworks and Project Management*. Stillwater: New Forums Press.

248. Shneiderman, B. 1992. Tree visualization with tree-maps: 2-d space-filling approach. *ACM Transactions on Graphics (TOG)* 11 (1): 92–99.

249. Sim, S.E., S. Easterbrook, and R.C. Holt. 2003. Using benchmarking to advance research: A challenge to software engineering. In *Proceedings of the 25th International Conference on Software Engineering*, 74–83. IEEE Computer Society.

250. Smith, H., and P. Fingar. 2003. *Business Process Management: The Third Wave*, vol. 1. Tampa: Meghan-Kiffer Press.
251. Spenader, J. 2012. Empirical approaches to discourse. In *Foundational Course at ESSLLI (European Summer School on Language, Logic and Information)*. Opole, Poland.
252. Stanley, R., and H. Astudillo. 2013. Ontology and semantic wiki for an intangible cultural heritage inventory, 1–12. In *CLEI*.
253. Stanley, R., H. Astudillo, V. Codocedo, and A. Napoli. 2013. A Conceptual-KDD approach and its application to cultural heritage. In *Concept Lattices and their Applications*, 163–174.
254. Stein, B., M. Koppel, and E. Stamatatos. 2007. Plagiarism analysis, authorship identification, and near-duplicate detection PAN'07. *SIGIR Forum* 41 (2): 68–71.
255. Stein, B., N. Lipka, and P. Prettenhofer. 2011. Intrinsic plagiarism analysis. *Language Resources and Evaluation* 45 (1): 63–82.
256. Stockinger, P. 1990. On Gardin's logicist analysis. *Interpretation in the Humanities: Perspectives from Artificial Intelligence*, 284–304. ed. *REeJ-CG*.
257. Svensson, P. 2009. Humanities computing as digital humanities. *Digital Humanities Quarterly* 6: 1.
258. Taboada, M. 2009. Implicit and explicit coherence relations. *Discourse, of Course*, 127–140. Amsterdam: John Benjamins.
259. Terras, M. 2005. Reading the readers: Modelling complex humanities processes to build cognitive systems. *Literary and Linguistic Computing* 20 (1): 41–59.
260. Torres-Moreno, J.-M. 2010. Reagrupamiento en familias y lexematización automática independientes del idioma. *Inteligencia Artificial* 47: 38–53.
261. Torres-Moreno, J.-M. 2012. Beyond stemming and lemmatization: Ultra-stemming to improve automatic text summarization. *arXiv preprint arXiv:1209.3126*.
262. Ulmer, J.-S., J.-P. Belaud, and J.-M.L. Lann. 2013. A pivotal-based approach for enterprise business process and IS integration. *Enterprise Information System* 7 (1): 61–78.
263. Unbehauen, J., S. Hellmann, S. Auer, and C. Stadler. 2012. Knowledge extraction from structured sources. In *Search Computing*, 34–52. Berlin: Springer.
264. UNESCO. 2003. *Convention for the Safeguarding of the Intangible Cultural Heritage*. http://unesdoc.unesco.org/images/0013/001325/132540e.pdf.
265. Unkelos-Shpigel, N., and I. Hadar. 2013. Using distributed cognition theory for analyzing the deployment architecture process. In *Advanced Information Systems Engineering Workshops*, 186–191. Berlin: Springer.
266. Unsworth, J. 2000. Scholarly primitives: What methods do humanities researchers have in common, and how might our tools reflect this. In *Humanities Computing, Formal Methods, Experimental Practice Symposium*, 5–100.
267. Vaishnavi, V., and W. Kuechler. 2004. Design Research in Information Systems. Association for Information Systems (AISWorld). Available on line at http://desrist.org/design-research-in-information-systems/ Association for Information Systems, 2005.
268. van der Linden, D., H.A. Proper, and S.J. Hoppenbrouwers. 2014. Conceptual understanding of conceptual modeling concepts: A longitudinal study among students learning to model. In *Advanced Information Systems Engineering Workshops*, 213–218. Berlin: Springer.
269. van Ham, F. 2012. Re-inventing and re-implementing the wheel. Visualization Component Reuse in a Large Enterprise, 5. In *SIGRAD*.
270. Van Lamsweerde, A. 2009. *Requirements Engineering: from System Goals to UML Models to Software Specifications*. West Sussex, England: John Wiley & Sons Ltd., ISBN-13: 978-0470012703.
271. Van Someren, M.W., Y.F. Barnard, and J.A. Sandberg. 1994. *The Think Aloud Method: A Practical Guide to Modelling Cognitive Processes*, vol. 2. London: Academic Press.
272. Van Veenendaal, E. 2012. Standard glossary of terms used in software testing. *International Software Testing Qualifications Board*, 8–9.

273. Van Velsen, L., T. Van der Geest, and R. Klaassen. 2007. Testing the usability of a personalized system: Comparing the use of interviews, questionnaires and thinking-aloud. In *Professional Communication Conference, 2007. IPCC 2007. IEEE International*, 1–8. IEEE.

274. van Waes, L. 2000. Thinking aloud as a method for testing the usability of websites: The influence of task variation on the evaluation of hypertext. *IEEE Transactions on Professional Communication* 43 (3): 279–291.

275. Vargas-Vera, M., E. Motta, J. Domingue, S.B. Shum, and M. Lanzoni. 2001. Knowledge extraction by using an ontology based annotation tool. In *Proceedings of the K-CAP 2001 Workshop on Knowledge Markup and Semantic Annotation*. Victoria, B.C., Canada.

276. von Alan, R.H., S.T. March, J. Park, and S. Ram. 2004. Design science in information systems research. *MIS Quarterly* 28 (1): 75–105.

277. Wang, Y. 2008. *Novel Approaches in Cognitive Informatics and Natural Intelligence*. IGI Global.

278. Warmer, J.B., and A.G. Kleppe. 1998. *The Object Constraint Language: Precise Modeling With UML*. Addison-Wesley Object Technology Series.

279. Weilkiens, T. 2008. *Systems Engineering with SysML/UML: Modeling, Analysis, Design*, 320. San Francisco: Morgan Kaufmann Publishers Inc.

280. Whyte, W.F. 2012. *Street Corner Society: The Social Structure of an Italian Slum*. Chicago: University of Chicago Press.

281. Wieringa, R. 2014. *Design Science Methodology for Information Systems and Software Engineering*, 3–317. Berlin: Springer.

282. Wieringa, R., and A. Moralı. 2012. Technical action research as a validation method in information systems design science. In *Design Science Research in Information Systems. Advances in Theory and Practice*, ed. K. Peffers, M. Rothenberger, and B. Kuechler. Lecture Notes in Computer Science, 220–238. Berlin, Heidelberg: Springer.

283. Wikipedia. 2015. *Software-Assisted Detection*. Accessed on 26/04/2016. http://en.wikipedia.org/wiki/Plagiarism_detection#Software-assisted_detection.

284. Wohlin, C., P. Runeson, M. Höst, M.C. Ohlsson, B. Regnell, and A. Wesslén. 2012. *Experimentation in Software Engineering*. Springer Science & Business Media.

285. Wolfe, M.B., and S.R. Goldman. 2005. Relations between adolescents' text processing and reasoning. *Cognition and Instruction* 23 (4): 467–502.

286. Wouters, P. 2006. What is the matter with e-Science?-thinking aloud about informatisation in knowledge creation, 23. In *Pantaneto forum*.

287. Wulf-Hadash, O., and I. Reinhartz-Berger. 2013. Constructing domain knowledge through cross product line analysis. In *Enterprise, Business-Process and Information Systems Modeling*, ed. S. Nurcan, et al. Lecture Notes in Business Information Processing, 354–369. Berlin, Heidelberg: Springer.

288. Yi, J.S., Y. ah Kang, J.T. Stasko, and J.A. Jacko. 2007. Toward a deeper understanding of the role of interaction in information visualization. *IEEE Transactions on Visualization and Computer Graphics* 13 (6): 1224–1231.

289. Yu, E.S. 1997. Towards modelling and reasoning support for early-phase requirements engineering. In *Proceedings of the Third IEEE International Symposium on Requirements Engineering*, 226–235. IEEE.

290. Zhang, L., B. Pan, W. Smith, and X.R. Li. 2009. An exploratory study of travelers' use of online reviews and recommendations. *Information Technology & Tourism* 11 (2): 157–167.

291. Zhou, M.X., and S.K. Feiner. 1998. Visual task characterization for automated visual discourse synthesis. In *Proceedings of the SIGCHI Conference on Human Factors in Computing Systems* 392–399. ACM Press/Addison-Wesley Publishing Co.

292. Zikra, I., J. Stirna, and J. Zdravkovic. 2011. Analyzing the integration between requirements and models in model driven development. In *Enterprise, Business-Process and Information Systems Modeling*, ed. T. Halpin, et al. Lecture Notes in Business Information Processing, 342–356. Berlin, Heidelberg: Springer.

Printed in the United States
By Bookmasters